Viral Interactions with Host RNA Decay Pathways

Special Issue Editors
J. Robert Hogg
Karen L. Beemon

MDPI • Basel • Beijing • Wuhan • Barcelona • Belgrade

MDPI

Special Issue Editors

J. Robert Hogg
National Institutes of Health
USA

Karen L. Beemon
The Johns Hopkins University
USA

Editorial Office
MDPI AG
St. Alban-Anlage 66
Basel, Switzerland

This edition is a reprint of the Special Issue published online in the open access journal *Viruses* (ISSN 1999-4915) from 2016–2017 (available at: http://www.mdpi.com/journal/viruses/special_issues/host_RNA_decay_pathway).

For citation purposes, cite each article independently as indicated on the article page online and as indicated below:

Author 1, Author 2. Article title. *Journal Name*. **Year**. Article number/page range.

First Edition 2017

ISBN 978-3-03842-502-1 (Pbk)
ISBN 978-3-03842-503-8 (PDF)

Table of Contents

About the Special Issue Editors

J. Robert Hogg Earl Stadtman Investigator, National Heart, Lung, and Blood Institute, National Institutes of Health. J. Robert Hogg graduated from Haverford College with a B.S. in Biology and earned a Ph.D. from the Department of Molecular and Cell Biology at the University of California, Berkeley. His thesis research in the laboratory of Kathleen Collins at UC Berkeley focused on the composition and function of ribonucleoprotein complexes containing noncoding RNAs. He pursued postdoctoral research with Stephen P. Goff at Columbia University, where he studied how the nonsense-mediated mRNA decay pathway selects its targets. He is currently an Earl Stadtman Investigator at the National Heart, Lung, and Blood Institute, National Institutes of Health. His laboratory studies mechanisms by which viral and cellular mRNAs evade cellular quality control.

Karen L. Beemon is a Professor of Biology, Johns Hopkins University. Dr. Beemon earned her B.S. in Cell Biology from the University of Michigan and her Ph.D. in Molecular Biology from the University of California, Berkeley. Her graduate work led to determination of the genomic complexity of retroviruses. She was the first postdoc of Tony Hunter at the Salk Institute, where she characterized the product of the SRC oncogene. At Hopkins, the Beemon lab has identified retroviral RNA regulatory elements, involved in post-transcriptional RNA processing steps, including RNA stability, splicing, nucleo-cytoplasmic export, and m6A methylation. She has also shown TERT and an associated antisense lnc RNA are often activated by ALV integration in chicken B-cell lymphomas. She served as Chair of the Department of Biology and Editor of the Journal of Virology.

Preface to "Viral Interactions with Host RNA Decay Pathways"

Eukaryotes have evolved a wide variety of RNA decay pathways to maintain cellular homeostasis, carry out programs of gene expression, and respond to changing environmental conditions. Individual RNA turnover mechanisms can operate constitutively or under only particular cellular conditions; similarly, some target many RNAs, while others act with great specificity. It has become increasingly clear that there are extensive interactions between viruses and the host RNA decay machinery. Often, the cellular RNA decay machinery poses a threat to viral gene expression, but viruses can also manipulate RNA decay pathways to promote viral replication. This special issue focuses on how cellular RNA decay factors recognize and degrade viral RNAs and viral strategies to subvert or evade these pathways.

J. Robert Hogg and Karen L. Beemon
Special Issue Editors

viruses

MDPI

Communication

Rous Sarcoma Virus RNA Stability Element Inhibits Deadenylation of mRNAs with Long 3′UTRs

Vidya Balagopal and Karen L. Beemon *

Biology Department, Johns Hopkins University, Baltimore, MD 21218, USA; vbalago1@jhu.edu
* Correspondence: KLB@jhu.edu; Tel.: +1-410-516-7289

Academic Editor: Eric O. Freed
Received: 5 June 2017; Accepted: 26 July 2017; Published: 1 August 2017

Abstract: All retroviruses use their full-length primary transcript as the major mRNA for Group-specific antigen (Gag) capsid proteins. This results in a long 3′ untranslated region (UTR) downstream of the termination codon. In the case of Rous sarcoma virus (RSV), there is a 7 kb 3′UTR downstream of the *gag* terminator, containing the *pol*, *env*, and *src* genes. mRNAs containing long 3′UTRs, like those with premature termination codons, are frequently recognized by the cellular nonsense-mediated mRNA decay (NMD) machinery and targeted for degradation. To prevent this, RSV has evolved an RNA stability element (RSE) in the RNA immediately downstream of the *gag* termination codon. This 400-nt RNA sequence stabilizes premature termination codons (PTCs) in *gag*. It also stabilizes globin mRNAs with long 3′UTRs, when placed downstream of the termination codon. It is not clear how the RSE stabilizes the mRNA and prevents decay. We show here that the presence of RSE inhibits deadenylation severely. In addition, the RSE also impairs decapping (DCP2) and 5′-3′ exonucleolytic (XRN1) function in knockdown experiments in human cells.

Keywords: Rous sarcoma virus; RNA stability element; nonsense mediated decay; long 3′UTR; deadenylation; decapping

1. Introduction

Retroviruses Have Aberrant mRNAs but Evade Cellular Nonsense-Mediated mRNA Decay

Retroviruses have compact genomes (<10 kilobases) that can code for eight or more viral proteins. In order to maximize the coding potential, major unspliced retroviral mRNAs often possess features such as long 3′ untranslated regions (UTRs), upstream open reading frames (uORFs), and retained introns that are predicted to be targets of host cell RNA surveillance machineries [1–3]. The unspliced RNA of Rous sarcoma virus (RSV), an avian retrovirus, carries a very long 3′UTR (ca. 7 kb); yet, it is very stable. A cis-acting RNA sequence found in the RSV genome, designated the RNA stability element (RSE; [4,5]) promotes viral evasion of the cellular nonsense-mediated mRNA decay NMD. The RSE, a 400-nt element located immediately downstream of the *gag* termination codon in the unspliced RSV viral RNA, has been shown to protect the viral RNA from Up-frameshift protein 1(UPF1) dependent decay in chicken cells [2,5–8]. Truncations in the RSE have defined a minimal RSE element of 155 nts [8].

More recently, the RSV RSE has been shown to protect cellular mRNAs with long 3′UTRs from NMD in mammalian cells [9]. Ge et al. inserted the RSE immediately downstream of the termination codon in a β-globin NMD reporter with a long SMG5 3′UTR, and found that it promoted stability of the reporter mRNA in human cells. Furthermore, the insertion of an antisense RSE fragment of similar length into this construct failed to stabilize it. The RSE RNA was shown to bind polypyrimidine tract binding protein 1 (PTBP1) and to decrease UPF1 association with the mRNA [9]. Interestingly, hundreds of cellular mRNAs with long 3′UTRs seem to have a similar PTBP1-binding sequence downstream of the termination codon that stabilizes the mRNAs [9]. However, the mechanism of mRNA stabilization by the RSE is unclear.

Degradation of NMD targets carrying premature termination codons (PTCs) has been shown to occur through exonucleolytic and endonucleolytic decay pathways [10,11]. mRNA decay is a multistep process regulated by several proteins. SMG6 has been shown to be important for endonucleolytic cleavage close to the PTC in some mRNAs [11]. Exonucleolytic degradation usually starts with the removal of the poly(A) tail (deadenylation). The deadenylated mRNA can then be decapped and subjected to 5′-3′ exonucleolytic decay. It can also become a substrate for the 3′-5′ exonucleases [10]. Accelerated deadenylation has been reported for PTC-containing NMD targets in mammalian cells [12,13]. However, the mechanism of decay of long 3′UTR-containing NMD targets is less well studied.

In this study, we asked how NMD targets containing long 3′UTRs are degraded and how the presence of the RSE allows them to evade NMD. We show that these NMD targets undergo deadenylation and decapping, as well as 5′-3′ exonucleolytic decay. Further, the presence of the RSE severely inhibits deadenylation and also impairs decapping and XRN1 mediated 5′-3′ exonucleolytic decay of the NMD reporter containing a long 3′UTR.

2. Materials and Methods

2.1. Cell Culture and Transfections

HeLa Tet-off Advance cells (Clonetech, Mountain View, CA, USA) were maintained in Dulbecco's modified eagle medium (DMEM) supplemented with 10% fetal bovine serum, 1X antibiotic-antimycotic (Gibco 15240-062, Gaithersburg, MD, USA), and 0.3 mg/mL L-Glutamine (Life Technologies, Carlsbad, CA, USA). Cells were transfected with desired constructs using FUGENE 6 transfection reagent (Promega, Madison, WI, USA) according to the manufacturer's protocol.

2.2. Analysis of Deadenylation and Decay

HeLa Tet-off Advance cells maintained in DMEM, supplemented with 10% fetal bovine serum (FBS) and 5 ng/mL doxycycline (Sigma, St. Louis, MO, USA), were seeded at a density of 1 million cells in 10-cm plates. For each plate, 4.8 μg of the indicated pcTET2-reporter plasmid was co-transfected with 1.2 μg of the wild-type β-globin reporter (βwt) control plasmid using FUGENE 6 transfection reagent (Promega) according to the manufacturer's instructions. Twenty-four hours post-transfection, cells were split into six equal aliquots in six-well plates. The next day, cells were washed twice with 2 mL 1X Phosphate Buffered Saline (PBS) and incubated in medium without doxycycline for 5 h. Transcription was shut off by adding doxycycline to a final concentration of 1 μg/mL, and cells were harvested in Trizol (Life Technologies) after 30 min (time 0) and at the indicated intervals. Fifteen micrograms of the total RNA were annealed to oligonucleotide OVB117 (5′-GAAAGTGATGCTTTAGTCTCAGTC-3′) complementary to a sequence that is 214 nucleotides upstream of the poly(A) addition site of the reporter mRNA to generate a shortened RNA form to measure poly(A) tail length. RNaseH/oligo treatment of mRNA to generate poly(A) minus mRNA and measure deadenylation and decay were carried out as described in Reference [14]. Gel mobility of the RNA bands were measured using the scale tool on adobe Photoshop and sizes were determined by comparison of gel mobility to known size of the various bands of the end labelled ladder used. Amount of DNA at each time point was quantitated using densitometry and compared to a co-transfected control. mRNA levels from at least three biological replicates performed using extracts from cells transfected separately were used in each experiment.

To detect mRNA levels in cells depleted for specific factors by RNA interference (RNAi) HeLa Tet-off cells were depleted for the proteins using RNAi plasmids. HeLa tet-off cells were seeded at 0.2 million cells into 6-well plates and grown in DMEM supplemented with 10% FBS. After 24 h, each well was transfected with 500 ng of the indicated pcTET2 reporter plasmid and 500 ng of RNAi plasmid using Fugene 6 transfection reagent. Doxycycline was added to a final concentration of 5 ng/mL. Two days post-transfection, the wells were washed with 1X PBS reagent twice and supplemented with media without doxycycline to turn on transcription of the reporter mRNA. Cells were collected in Trizol after 12 h.

2.3. Northern Blotting

RNA was isolated using Trizol and resolved on 1.4% formaldehyde/agarose gels. ^{32}P end-labeled oligonucleotides (GGAGTGGCACCTTCCAGGGTCAAG) against bovine growth hormone (bGH) polyadenylation signal present in both βwt and Tet-off reporter constructs were used for the detection of mRNAs. Northern blots were imaged on Typhoon phosphor imager scanners and quantification was performed using ImageQuant software (GE Healthcare Bio-Sciences Corp, Piscataway, NJ, USA).

3. Results

3.1. The RSV RSE Impairs mRNA Deadenylation

We have previously shown that the RSE can inhibit NMD of mRNAs with long 3′UTRs [5,9]. We wanted to understand how the RSE affects the mechanism of decay. Since deadenylation is often the first step in decay, we first studied the deadenylation rates of NMD reporters with long 3′UTRs in the presence or absence of the RSE. To address this question, we conducted a transcriptional-pulse chase experiment to determine the deadenylation and decay rates. Here, we used two reporter constructs, WT-SMG5 and RSE-SMG5, under the control of a Tet-off promoter ([9], Figure 1). The tetracycline (tet)-regulated reporter WT-SMG5 mRNA contains a β-globin mini-gene (with introns) and the SMG5 3′UTR. The SMG5 3′UTR has been shown to trigger NMD; this is proposed to be due to its length (1342 nt) [15–17]. The RSE-SMG5 construct carries a 400-nt RSE sequence inserted immediately downstream of the stop codon, mimicking the natural context of the RSE in the RSV RNA.

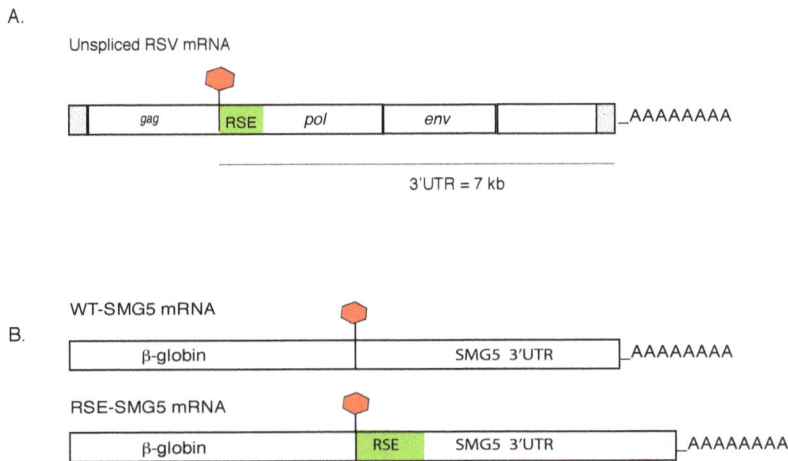

Figure 1. (**A**) Schematic of the Rous sarcoma virus (RSV) unspliced mRNA. The 400-nt RNA stability element (RSE) (shown in green) is located immediately downstream of the *gag* stop codon. This RNA has a very long 3′ untranslated region (UTR) of ca. 7 kb. (**B**) Schematic of tet-regulated β-globin reporter mRNA constructs used in our studies. WT-SMG5 mRNA contains the β-globin sequence followed by the SMG5 3′UTR sequence. The RSV RSE sequence was inserted into this reporter construct immediately after the β-globin stop codon to generate RSE-SMG5.

These reporter constructs were co-transfected into HeLa tet-off cells (Clonetech, Mountain View, CA, USA) with a vector constitutively expressing a control RNA. The expression of tet-regulated mRNAs was induced for 5 h before transcription was shut off by the addition of doxycycline, and mRNA was collected at the indicated time points. The mRNAs were then subjected to oligo-dT /RNaseH treatments followed by agarose Northern analysis to measure the poly(A) tail length and/or decay over the time course (Figure 2). At the starting time, both constructs generated mRNAs with poly(A) tails of approximately 200 nts (Figure 2B).

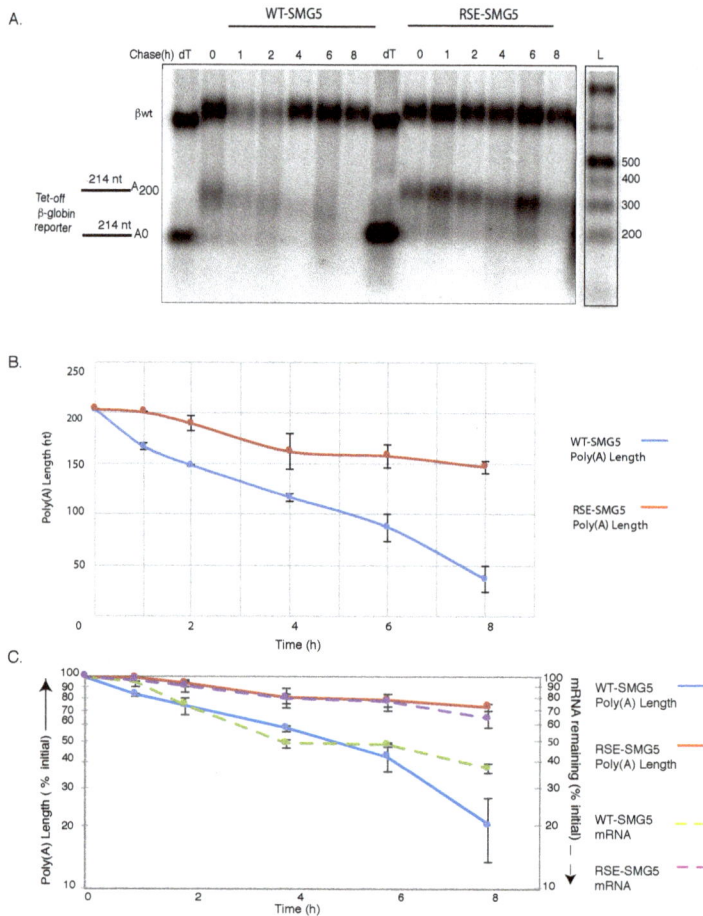

Figure 2. RNA stability element inhibits deadenylation. (**A**) Agarose northern blot showing deadenylation and decay of WT-SMG5 and RSE-SMG5 mRNAs following tet-off transcriptional pulse chase. RNA was subjected to oligonucleotide-directed RNaseH treatment prior to Northern blotting for tail length determination. Lanes marked "dT" show the length of the RNA fragment without a poly(A) tail; 1 kb plus DNA ladder (ThermoFisher Catalog number: 10787018, Halethorpe, MD, USA) was end labelled and used for size determination (size corrected for RNA). Wildtype β-globin (βwt) (upper band) was used as a loading control. (**B**) Poly(A) tail length determined from the gel was plotted against time to show the initial and final tail length of the two reporter constructs. The plotted values are the average of at least three separate experiments. Error bars show the standard error. Deadenylation rate of the reporters were also determined from this analysis. (**C**) Deadenylation and decay of the reporter mRNAs were compared by plotting mRNA remaining (% of original) and poly(A) length remaining (% of original tail length) on the same plot.

The WT-SMG5 reporter underwent rapid deadenylation at the rate of ca. 0.5 A/min, while the RSE-SMG5 reporter showed deadenylation rates that slowed down to ca. 0.1 A/min (Figure 2B). Deadenylation in mammalian cells usually proceeds in three steps. A slow initial deadenylation, usually by the weaker deadenylase complex Pan 2/3, results in ca. 110-nt A tails. This is followed by CCR4-NOT complex-mediated deadenylation resulting in a tail of ca. 22 As. In some cases, a terminal deadenylation occurs where all the As are removed [12]. We observed that the deadenylation of the RSE containing constructs did not surpass 70% of the original tail length in our 8-h time course. In contrast, the WT-SMG5 mRNA deadenylated to a poly(A) tail of <50 nts in 8 h (Figure 2B,C).

We compared the deadenylation and decay rates of WT-SMG5 and RSE-SMG5 reporters (Figure 2C). Transcripts containing only the SMG5 3'UTR exhibited a half-life of ca. 200 min while the transcripts containing the RSE were substantially more stable (half-life > 480 min), confirming the protective activity of the RSE in agreement with previous reports [9]. It is interesting to note that the decay rates of each reporter construct closely mirrored their deadenylation rates. This would suggest that the change in deadenylation is responsible for most of the alteration in decay.

3.2. Effects of the RSE on Decapping and 5'→3' Exonucleolytic Decay

Nonsense mediated decay of mRNAs with long 3'UTRs could be exonucleolytic, initiating at either or both mRNA ends, or endonucleolytic. We assessed the role of the RSE in the regulation of decapping and 5'→3' exonucleolytic decay. Decapping mRNA 2 (DCP2) or 5'-3' Exoribonuclease 1 (XRN1) were depleted in HeLa-tet off cells using pSUPuro-Based RNAi plasmids previously characterized in Eberle et al. (generous gift from Oliver Muhlemann) [11]. Knockdown was performed by co-transfecting reporter constructs with pSUPuro-based RNAi plasmids against DCP2 or XRN1. We then measured the accumulation of the reporter constructs at steady state. As a control, RNAi against a scrambled sequence was separately carried out.

Upon depletion of the major decapping enzyme DCP2, we found the WT-SMG5 reporter construct to be stabilized ca. 2-fold. The RSE-SMG5 construct was stabilized ca. 1.6-fold in this knockdown, as compared to the corresponding scrambled controls. Depletion of XRN1, the major 5'-3' exonuclease, showed even greater accumulation of ca. 4.3-fold in WT-SMG5 and ca. 2.5-fold in RSE-SMG5 (Figure 3). Thus, impairment of decay by down regulating DCP2 or XRN1 had less effect on the RSE-containing construct than on the wildtype construct. This suggests the RSE is protecting the mRNA from degradation at the 5' end of the mRNA as well as at the 3' end. We cannot tell from these experiments whether the deadenylation precedes the decapping and 5'→3' exonuclease activities.

We also noticed that XRN1 had a stronger effect on the accumulation of mRNA than DCP2 in both WT-SMG5 and RSE-SMG5 constructs. Knockdown of XRN1 mRNA appears to be more robust than DCP2 (Figure 3C). This could also be because there may be redundant decapping enzymes present in the cell such as NUDT16 [18,19]. We also observed that DCP2 and XRN1 depletions led to stabilizations of both reporter constructs. It is not surprising that the RSE-SMG5 reporter was stabilized upon depletion of both DCP2 and XRN1, as both proteins are components of the general decay pathway and are important for the decay of most normal messages. It should be noted that the WT-SMG5 constructs were always more stabilized than the RSE-SMG5 in both knockdowns.

It has been previously reported that the depletion of XRN1 leads to the accumulation of a 3' mRNA fragment for NMD targets that is generated by endonucleolytic cleavage [11]. We were unable to detect the presence of 3' fragments in our experiments. This would suggest that our NMD reporter does not undergo decay via endonucleolytic cleavage. It is, of course, possible that endonucleolytic cleavage is very sensitive to minor changes in conditions.

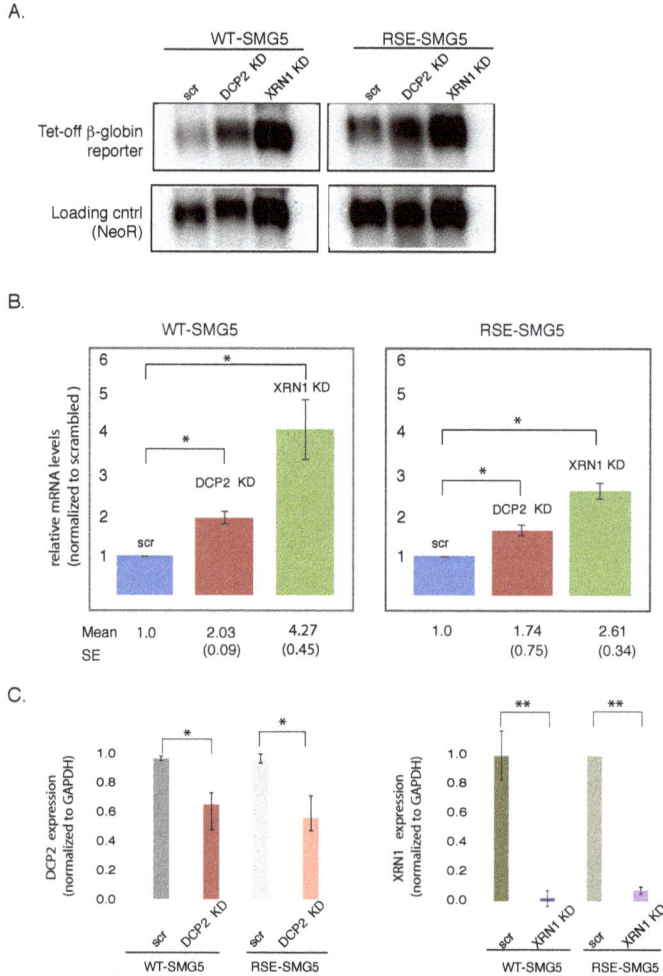

Figure 3. Depletion of decapping mRNA 2 (DCP2) or 5′-3′ Exoribonuclease 1(XRN1) leads to the accumulation of reporter mRNAs. (**A**) mRNA levels of the two reporters isolated from cells depleted for Exoribonuclease 1 (XRN1 KD), decapping mRNA 2 (DCP2 KD), or scrambled control (scr) were detected by agarose Northern blotting. NeoR, an mRNA produced from the same plasmid that generates the tet-off β-globin constructs was used as a loading control. Mean numbers show the amount of mRNA relative to scrambled control for WT-SMG5. SE shows the standard error of the values. (**B**) Left panel shows levels of WT-SMG5 reporter in different knockdown conditions normalized to the scrambled control. Right panel shows levels of RSE-SMG5 reporter in different knockdown conditions normalized to the corresponding scrambled control. The plotted values are the average of at least three separate experiments. Error bars show the standard error. (**C**) Quantitative reverse transcription polymerase chain reaction (RT-PCR) was carried out to determine the levels of knockdown of DCP2 and XRN1. Expression levels of these mRNAs were measured relative to the housekeeping gene glyceraldehyde 3-phosphate dehydrogenase (GAPDH). Unpaired *t*-test was used to test for significance of the change in expression between each knockdown and the corresponding scrambled control. * denotes *p*-value < 0.05. ** *p*-value < 0.01.

4. Discussion

mRNA degradation can be initiated by generating a new unprotected end vulnerable to attack by an exonuclease. In human cells, unprotected ends can be generated by endonucleolytic cleavage, deadenylation-dependent decapping, and exosome-mediated 3'-to-5' decay. Deadenylation is one of the slowest and often rate-limiting steps in degradation. In this study, we looked at how the RSE protects long 3'UTR-containing mRNAs from NMD.

Our results show that the presence of the RSE severely inhibits deadenylation. We measured the tail length of the WT globin-SMG5 reporter to be ca. 200 As. The deadenylation rates of this NMD target mRNA was determined to be ca. 0.5 A/min. Both tail length and deadenylation rates are in agreement with previous studies for PTC-containing globin NMD reporters [14]. In the presence of the RSE, the WT-SMG5 reporter maintained the initial tail length of 200 As, but the deadenylation rate slowed down to approximated 0.1 A/min. This five-fold change in deadenylation is a significant impairment and could be a leading causal factor in the impairment of decay. It is also interesting to note that the deadenylation of the WT-SMG5 reporter proceeds to <50 As. The length of the poly(A) tail is known to affect translation and decay rates. mRNAs with short poly(A) tails (<50 A residues) are generally translationally repressed (reviewed in [20]).

We compared the deadenylation and decay rates to estimate the contribution of deadenylation to the degradation of the reporters. In both WT-SMG5 and RSE-SMG5, the decay rates follow the deadenylation rates closely. This is different from the case for the PTC-containing β-globin reporter, which undergoes rapid deadenylation but the decay rates seem to lag behind [13]. Our observation suggests that deadenylation plays an important role in the degradation of long 3'UTR-containing NMD targets and that the protective function of the RSE works through inhibiting deadenylation. How the RSE regulates deadenylation remains an interesting open question. We have previously shown that RSE can inhibit UPF1 binding to the RSE-SMG5 reporter mRNA. The UPF proteins, the SMG proteins, and the eRF1-eRF3 complex constitute the SURF complex that is important for triggering NMD in PTC-containing mRNA [21].

Nonsense mediated decay can be prevented by placing cytoplasmic poly(A) binding protein (PABP) in proximity to the termination codon, suggesting that the increased distance between the translation termination event and cytoplasmic PABP that results from termination at a PTC contributes to NMD [11,16,22–24]. UPF1 has also been shown to play an important role in accelerated decay in PTC-containing mRNAs [13]. This could be via interactions with SMG7 that has been shown to recruit the CCR4-NOT deadenylation complex by directly binding to POP2, its catalytic subunit [25]. The RSE could disrupt UPF1 binding, reducing SMG7 binding and hence the recruitment of the deadenylation complex. It has been reported that the degradation activity of SMG7 involves the decapping enzyme DCP2 and the 5'-to-3' exonuclease XRN1 [25].

Using RNAi knockdown experiments, we show that the depletion of DCP2 and XRN1 results in the enhanced steady-state accumulation of mRNA with both reporter constructs. This is not surprising, as DCP2 and XRN1 are part of the general mRNA decay pathway. The effect of depletion of these factors is more pronounced in the WT-SMG5 NMD reporter mRNA. Since deadenylation, decapping, and XRN1 regulate the decay of the long 3'UTR-containing NMD target, we speculate that SMG7 might play an important role in the decay. Our results are in agreement with previous results with other NMD substrates. Lejeune and Maquat have shown the importance of decapping and 5'-3' decay, as well as deadenylation, in the degradation of PTC-containing mRNA [10].

Given that poly(A) shortening is often the first and rate-limiting step in mRNA decay, viruses likely have developed cis and trans acting factors to repress or circumvent deadenylation [26]. Several families of RNA viruses, such as flaviviruses, bunyaviruses, and arenaviruses, have evolved 3' terminal stem loop structures to stabilize the RNA in the absence of poly(A) tails [27]. Poliovirus infection promotes the degradation of poly(A) specific ribonuclease subunit 3 (PAN3), a protein that initiates the deadenylation of many cellular mRNAs [28]. Sindbis virus recruits the cellular HuR protein to stability elements in the 3'UTR of its transcripts to stabilize its poly(A) tail [29,30]. Kaposi's sarcoma associated herpesvirus (KSHV) PAN RNA has a 79-nucleotide expression and nuclear retention element (ENE) that forms a triple helix structure with the poly(A) tail that inhibits degradation [31]. Our results using reporter constructs suggest that the Rous sarcoma virus RSE in its natural context functions to prevent NMD by inhibiting deadenylation, and thus adds to the growing number of examples of viruses manipulating deadenylation to escape cellular decay machinery.

Acknowledgments: This work was supported by National Institutes of Health (NIH) RO1 CA048746. We thank Bobby Hogg for the globin expression constructs. We are grateful to Oliver Muhlemann for the constructs used to knock down DCP2 and XRN1. We thank all members of the Beemon lab, especially Yingying Li, for technical assistance and helpful discussions.

Author Contributions: V.B. carried out the experiments. V.B. and K.L.B. designed the experiments and wrote the manuscript.

Conflicts of Interest: The authors declare no conflict of interest.

References

1. Bolinger, C.; Boris-Lawrie, K. Mechanisms employed by retroviruses to exploit host factors for translational control of a complicated proteome. *Retrovirology* **2009**, *6*, 8. [CrossRef] [PubMed]
2. Withers, J.B.; Beemon, K.L. The structure and function of the Rous sarcoma virus RNA stability element. *J. Cell. Biochem.* **2011**. [CrossRef] [PubMed]
3. Leblanc, J.; Weil, J.; Beemon, K. Posttranscriptional regulation of retroviral gene expression: Primary RNA transcripts play three roles as pre-mRNA, mRNA, and genomic RNA. *Wiley Interdiscip. Rev. RNA* **2013**, *4*, 567–580. [CrossRef] [PubMed]
4. Barker, G.F.; Beemon, K. Rous sarcoma virus RNA stability requires an open reading frame in the gag gene and sequences downstream of the gag-pol junction. *Mol. Cell. Biol.* **1994**, *14*, 1986–1996. [CrossRef] [PubMed]
5. Weil, J.; Beemon, K. A 3′UTR sequence stabilizes termination codons in the unspliced RNA of Rous sarcoma virus. *RNA* **2006**, *12*, 102–110. [CrossRef] [PubMed]
6. Quek, B.L.; Beemon, K. Retroviral strategy to stabilize viral RNA. *Curr. Opin. Microbiol.* **2014**, *18*, 78–82. [CrossRef] [PubMed]
7. Weil, J.E.; Hadjithomas, M.; Beemon, K.L. Structural characterization of the Rous sarcoma virus RNA stability element. *J. Virol.* **2009**, *83*, 2119–2129. [CrossRef] [PubMed]
8. Withers, J.B.; Beemon, K.L. Structural features in the Rous sarcoma virus RNA stability element are necessary for sensing the correct termination codon. *Retrovirology* **2010**, *7*, 65. [CrossRef] [PubMed]
9. Ge, Z.; Quek, B.L.; Beemon, K.L.; Hogg, J.R. Polypyrimidine tract binding protein 1 protects mRNAs from recognition by the nonsense-mediated mRNA decay pathway. *eLife* **2016**, *5*. [CrossRef] [PubMed]
10. Lejeune, F.; Li, X.; Maquat, L.E. Nonsense-mediated mRNA decay in mammalian cells involves decapping, deadenylating, and exonucleolytic activities. *Mol. Cell* **2003**, *12*, 675–687. [CrossRef]
11. Eberle, A.B.; Lykke-Andersen, S.; Mühlemann, O.; Jensen, T.H. SMG6 promotes endonucleolytic cleavage of nonsense mRNA in human cells. *Nat. Struct. Mol. Biol.* **2009**, *16*, 49–55. [CrossRef] [PubMed]
12. Chen, C.Y.A.; Shyu, A. Bin Mechanisms of deadenylation-dependent decay. *Wiley Interdiscip. Rev. RNA* **2011**, *2*, 167–183. [CrossRef] [PubMed]
13. Chen, C.-Y.A.; Shyu, A.-B. Rapid deadenylation triggered by a nonsense codon precedes decay of the RNA body in a mammalian cytoplasmic nonsense-mediated decay pathway. *Mol. Cell. Biol.* **2003**, *23*, 4805–4813. [CrossRef] [PubMed]
14. Schiavi, S.C.; Wellington, C.L.; Shyu, A.B.; Chen, C.Y.A.; Greenberg, M.E.; Belasco, J.G. Multiple elements in the *c-fos* protein-coding region facilitate mRNA deadenylation and decay by a mechanism coupled to translation. *J. Biol. Chem.* **1994**, *269*, 3441–3448. [PubMed]
15. Huang, L.; Lou, C.H.; Chan, W.; Shum, E.Y.; Shao, A.; Stone, E.; Karam, R.; Song, H.W.; Wilkinson, M.F. RNA homeostasis governed by cell type-specific and branched feedback loops acting on NMD. *Mol. Cell* **2011**, *43*, 950–961. [CrossRef] [PubMed]
16. Singh, G.; Rebbapragada, I.; Lykke-Andersen, J. A competition between stimulators and antagonists of Upf complex recruitment governs human nonsense-mediated mRNA decay. *PLoS Biol.* **2008**, *6*, 860–871. [CrossRef] [PubMed]

17. Yepiskoposyan, H.; Aeschimann, F.; Nilsson, D.; Okoniewski, M.; Mühlemann, O. Autoregulation of the nonsense-mediated mRNA decay pathway in human cells. *RNA* **2011**, *17*, 2108–2118. [CrossRef] [PubMed]
18. Song, M.G.; Li, Y.; Kiledjian, M. Multiple mRNA decapping enzymes in mammalian cells. *Mol. Cell* **2010**, *40*, 423–432. [CrossRef] [PubMed]
19. Song, M.-G.; Bail, S.; Kiledjian, M. Multiple Nudix family proteins possess mRNA decapping activity. *RNA* **2013**, *19*, 390–399. [CrossRef] [PubMed]
20. Pichon, X.A.; Wilson, L.; Stoneley, M.; Bastide, A.A.; King, H.; Somers, J.E.; Willis, A. RNA binding protein/RNA element interactions and the control of translation. *Curr. Protein Pept. Sci.* **2012**, *13*, 294–304. [CrossRef] [PubMed]
21. Peixeiro, I.; Silva, A.L.; Romão, L. Control of human β-globin mRNA stability and its impact on beta-thalassemia phenotype. *Haematologica* **2011**, *96*, 905–913. [CrossRef] [PubMed]
22. Amrani, N.; Ganesan, R.; Kervestin, S.; Mangus, D.A.; Ghosh, S.; Jacobson, A. A faux 3′-UTR promotes aberrant termination and triggers nonsense-mediated mRNA decay. *Nature* **2004**, *432*, 112–118. [CrossRef] [PubMed]
23. Silva, A.L.; Romão, L. The mammalian nonsense-mediated mRNA decay pathway: To decay or not to decay! Which players make the decision? *FEBS Lett.* **2009**, *583*, 499–505. [CrossRef] [PubMed]
24. Fatscher, T.; Boehm, V.; Weiche, B.; Gehring, N.H. The interaction of cytoplasmic poly(A)-binding protein with eukaryotic initiation factor 4G suppresses nonsense-mediated mRNA decay. *RNA* **2014**, *20*, 1579–1592. [CrossRef] [PubMed]
25. Loh, B.; Jonas, S.; Izaurralde, E. The SMG5-SMG7 heterodimer directly recruits the CCR4-NOT deadenylase complex to mRNAs containing nonsense codons via interaction with POP$_2$. *Genes Dev.* **2013**, *27*, 2125–2138. [CrossRef] [PubMed]
26. Dickson, A.M.; Wilusz, J. Strategies for viral RNA stability: Live long and prosper. *Trends Genet.* **2011**, *27*, 286–293. [CrossRef] [PubMed]
27. Moon, S.L.; Barnhart, M.D.; Wilusz, J. Inhibition and avoidance of mRNA degradation by RNA viruses. *Curr. Opin. Microbiol.* **2012**, *15*, 500–505. [CrossRef] [PubMed]
28. Dougherty, J.D.; White, J.P.; Lloyd, R.E. Poliovirus-mediated disruption of cytoplasmic processing bodies. *J. Virol.* **2011**, *85*, 64–75. [CrossRef] [PubMed]
29. Garneau, N.L.; Sokoloski, K.J.; Opyrchal, M.; Neff, C.P.; Wilusz, C.J.; Wilusz, J. The 3′ untranslated region of sindbis virus represses deadenylation of viral transcripts in mosquito and mammalian cells. *J. Virol.* **2008**, *82*, 880–892. [CrossRef] [PubMed]
30. Sokoloski, K.J.; Dickson, A.M.; Chaskey, E.L.; Garneau, N.L.; Wilusz, C.J.; Wilusz, J. Sindbis virus usurps the cellular HuR protein to stabilize its transcripts and promote productive infections in mammalian and mosquito cells. *Cell Host Microbe* **2010**, *8*, 196–207. [CrossRef] [PubMed]
31. Mitton-Fry, R.M.; DeGregorio, S.J.; Wang, J.; Steitz, T.A.; Steitz, J.A. Poly(A) tail recognition by a viral RNA element through assembly of a triple helix. *Science* **2010**, *330*, 1244–1247. [CrossRef] [PubMed]

viruses

MDPI

Review

Virus Escape and Manipulation of Cellular Nonsense-Mediated mRNA Decay

Giuseppe Balistreri [1,*], Claudia Bognanni [2,3] and Oliver Mühlemann [2,*]

1   Department of Biosciences, University of Helsinki, Helsinki FIN-00014, Finland
2   Department of Chemistry and Biochemistry, University of Bern, Bern CH-3012, Switzerland;
     claudia.bognanni@dcb.unibe.ch
3   Graduate School for Cellular and Biomedical Sciences, University of Bern, Bern CH-3012, Switzerland
*   Correspondence: giuseppe.balistreri@helsinki.fi (G.B.); oliver.muehlemann@dcb.unibe.ch (O.M.);
     Tel.: +358-294159550 (G.B.); +41-31-631-4627 (O.M.)

Academic Editors: Robert Hogg and Karen L. Beemon
Received: 18 November 2016; Accepted: 13 January 2017; Published: 23 January 2017

Abstract: Nonsense-mediated mRNA decay (NMD), a cellular RNA turnover pathway targeting RNAs with features resulting in aberrant translation termination, has recently been found to restrict the replication of positive-stranded RNA ((+)RNA) viruses. As for every other antiviral immune system, there is also evidence of viruses interfering with and modulating NMD to their own advantage. This review will discuss our current understanding of why and how NMD targets viral RNAs, and elaborate counter-defense strategies viruses utilize to escape NMD.

Keywords: RNA quality control; gene expression; translation; RNA-protein interactions

1. Introduction

Virus Infections and Cellular mRNA Quality Controls

For their efficient replication, viruses rely on the host cell's replicative machinery and must avoid being recognized by cellular antiviral responses. Having co-evolved with their hosts for millions of years, viruses have developed means to modulate cellular functions and redirect the metabolic resources of the infected cell to their advantage. Antiviral defense mechanisms, on the other hand, have arisen to counteract pathogen infections. Recent evidence indicates that cellular RNA quality control systems, such as the nonsense-mediated mRNA decay (NMD) pathway, can restrict viral infections by different mechanisms [1,2], suggesting that the cellular RNA decay machinery could act as an ancestral form of intrinsic antiviral immunity [3]. Supporting this hypothesis, there is increasing evidence that viruses have evolved mechanisms to interfere with or modulate NMD at different post-entry stages of their replication cycle. In this review, we will analyze current evidence that supports a role of NMD in counteracting virus infections. To set the stage, we first describe the known features that make a cellular mRNA a substrate for NMD. Next, we apply this knowledge on viral RNAs and discuss how viral mRNAs could be recognized and attacked by the NMD surveillance machinery. Additionally, we will discuss which counter-defense strategies viruses utilize to ensure stability of their transcripts.

2. The Nonsense-Mediated mRNA Decay (NMD) Pathway

Eukaryotic cells possess numerous RNA quality control systems that degrade faulty mRNAs and so prevent the production of aberrant proteins. These mechanisms, including the NMD, the non-stop mRNA decay (NSD), and the no-go decay (NGD) pathways, are important for dynamically shaping the transcriptome and the proteome of eukaryotic cells to variable physiological conditions [1].

Of these mechanisms, NMD is arguably the best characterized [2]. NMD modulates the RNA levels of about 10% of all genes in diverse eukaryotes [3,4].

NMD was originally identified as an mRNA degradation pathway targeting transcripts that harbor a premature termination codon (PTC) [5–7]. However, in the last decade, it has become clear that NMD targets include a much broader number of substrates with different features. Transcripts targeted by NMD include those with PTCs in an internal exon or sometimes also in the terminal exon, as well as transcripts with long 3'-untranslated regions (UTRs), introns downstream a termination codon (TC) or upstream open reading frames (uORFs; Figure 1) [8–11]. The molecular details of how NMD is regulated and how NMD factors are recruited to target transcripts are not yet fully understood. It is clear that translation is a prerequisite for NMD and converging evidence suggests that a failure to terminate translation correctly triggers NMD [2]. This implies that any RNA molecule that engages with ribosomes to undergo translation is a potential target for NMD, including viral RNAs.

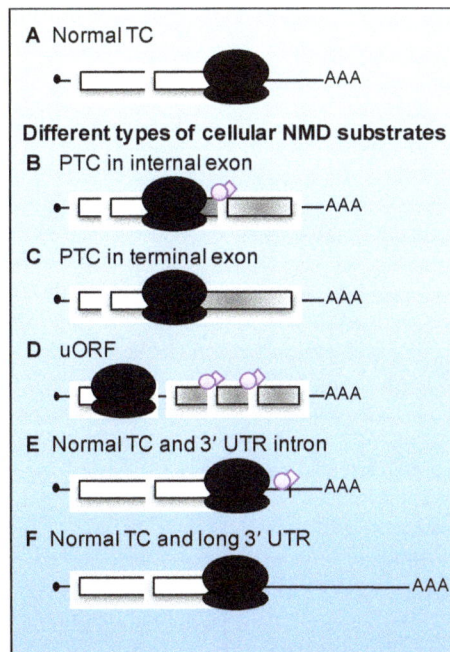

Figure 1. Different types of cellular mRNAs that can be substrates for nonsense-mediated mRNA decay (NMD). (**A**) An mRNA with a normal termination codon (TC) positioned in a context that does not trigger NMD: the termination codon is at the end of the last exon followed by a short 3'-untranslated region (3'-UTR). (**B–F**) NMD targets comprise mRNAs with a truncated ORF due to a premature termination codon (PTC) in an internal (**B**) or terminal (**C**) exon. The presence of protein complexes known as exon junction complexes (EJCs) downstream the TC increases, but it is not necessary for, NMD (see Section 3). Upstream ORF (uORF), (**D**), the presence of EJC-associated introns in the 3'-UTR (**E**), and long 3'-UTR (**F**), all act as RNA destabilizing factors and trigger NMD. White boxes denote translated ORFs; gray boxes denote the fraction of the ORF that is not translated due to the presence of a PTC. Ribosomes are indicated in black, EJC in purple.

3. NMD Factors

As detailed in several recent reviews, the activation of NMD requires a set of evolutionarily conserved core regulatory factors called up-frameshift (UPF) 1 to 3 and additional factors in metazoans [2,11–13].

UPF1 is a monomeric superfamily 1 (SF1) helicase that is essential for substrate recognition and NMD execution. The helicase domain consists of two RecA domains that bind single-stranded RNA (ssRNA) and DNA (ssDNA). It is flanked by an N-terminal cysteine and histidine rich (CH) domain that binds UPF2 [14], the ribosomal protein RPS26 [15] and the decapping enzyme subunit DCP2 [16], and by an unstructured C-terminal domain enriched in serine-glutamine (SQ) dipeptides.

UPF2, the second conserved NMD factor, functions as a scaffold linking UPF1 and UPF3. Human UPF2 has three tandem MIF4G domains and a C-terminal UPF1 binding region. MIF4G-3 interacts with UPF3 while the specific functions of MIF4G-1 and MIF4G-2 are unknown [17]. Besides bridging UPF1 with UPF3, UPF2 functions as an activator of UPF1 by promoting its helicase activity, thereby switching UPF1 from the RNA clamping to the RNA unwinding mode [18].

Of the three UPF proteins, UPF3 is the least conserved one [7]. Two UPF3 versions encoded by two different genes exist in mammals, UPF3A and UPF3B. Both are predominantly nuclear and shuttle between the nucleus and the cytoplasm. They are associated with spliced mRNA and bind UPF2 and the exon junction complex (EJC) [19], a multi protein complex comprising four core components (eIF4A1, MAGOH, Y14 and MLN51) and more than a dozen other factors [20], which upon splicing assembles 24 nucleotides upstream of the spliced junction and remains associated with the mRNA until translation, thereby "marking" the exon-exon junction.

Tethering experiments showed that UPF3A is less active than UPF3B in promoting NMD [21]. In human cells, UPF3B is the predominantly expressed protein and UPF3A becomes specifically upregulated if UPF3B levels are experimentally decreased or low because of a mutation [22]. Interestingly, a recent study shows that UPF3A is critical for spermatogenesis and provides evidence that it can antagonize NMD by sequestering UPF2, while its paralog, UPF3B activates NMD [23].

In addition to the three UPF factors, activation of NMD in metazoans requires the function of several additional proteins, among them a set of proteins originally identified in *Caenorhabditis elegans* known as the "suppressor with morphogenetic defects in genitalia" (SMG), SMG1 and SMG 5–9. These proteins control the phosphorylation and de-phosphorylation of UPF1 and trigger mRNA degradation by recruiting specific mRNA decay activities.

SMG1 belongs to the phosphatidylinositol 3 kinase-related protein kinase (PIKK) super family and it is responsible for phosphorylation of UPF1 at multiple S/TQ motifs [24]. SMG1, SMG8 and SMG9 form together with hypophosphorylated UPF1 and the eukaryotic release factors (eRF) 1 and 3 a complex called SURF [25,26], in which SMG8 and SMG9 repress the kinase activity of SMG1 [27]. Upon dissociation of SMG8 and SMG9 from SMG1, UPF2 mediates a structural rearrangement between SMG1 and UPF1 that activates the kinase, leading to UPF1 phosphorylation [28].

SMG5, SMG6 and SMG7 interact preferentially with hyperphosphorylated UPF1 and hence function further downstream in the NMD pathway. SMG5 and SMG7 form a heterodimer which interacts with phosphorylated SQ motifs in the C-terminal part of UPF1 and which recruits the CCR4-NOT deadenylase complex onto the mRNA [29,30]. SMG6 by contrast appears to function as a monomer [31]. Its C-terminal PIN domain has endonuclease activity [32] and it interacts with UPF1 through phosphorylated T28 as well as phosphorylation-independent contacts in the helicase domain [33,34]. In addition to the abovementioned NMD factors, additional proteins have more recently been shown to be involved in NMD and their molecular function is currently being investigated [25,35–37].

4. Current NMD Model

Aberrant Translation Termination Activates NMD

During normal translation termination, the ribosome when arriving at a termination codon (TC) binds eRF1 and eRF3. eRF1 recognizes the TC in the A-site of the stalling ribosome and forms a complex with the GTPase eRF3 to catalyze peptide release [38]. After eRF3-mediated GTP hydrolysis, eRF1 interacts with the ATP-binding cassette subunit family E member 1 (ABCE1). This interaction induces a structural change that stimulates ATP hydrolysis and leads to the separation of the two ribosomal subunits and the mRNA [39,40]. Importantly, in the context of NMD, the cytoplasmic polyA binding protein 1 (PABPC1) has recently been shown to enhance the recruitment of the eRFs, thereby stimulating correct and efficient translation termination [41]. Most likely due to this translation termination promoting activity, PABPC1 can potently antagonize NMD [42,43]. Based on these observations, it was postulated that whether an mRNA is subjected to NMD mainly depends on a competition between UPF1 and PABPC1 for binding to eRF3 [44]. The outcome of this competition is predicted to depend on the location of UPF1 and PABPC1 binding sites relative to the TC. Consistent with this model, TCs located in an mRNA towards the 3' end in the vicinity of the polyA tail (i.e., PABPC1 binding sites) will usually not elicit NMD. In contrast, PTCs, which can occur anywhere in the mRNA, as well as long 3' UTRs may allow UPF1 rather than PABPC1 to interact with eRF3 at the terminating ribosome and as a consequence activate NMD by formation of the SURF complex and subsequent UPF1 phosphorylation. Alternatively, if a TC is located more than 50 nucleotides upstream of the final exon–exon junction on the mRNA, an EJC most likely remains bound at this exon-exon junction, a constellation that is well known to activate NMD [45]. It is thought that in this situation the EJC-bound UPF3B, via UPF2, recruits UPF1 and promotes SURF complex formation and NMD activation.

The observation that ribosomes reside longer at TCs on transcripts subject to NMD [46,47] indicates that NMD-triggering translation termination is mechanistically distinct from proper termination. However, the mechanistic differences between proper and NMD-triggering termination have remained elusive so far.

After formation of the SURF complex on an NMD targeted mRNA and subsequent SMG1-mediated phosphorylation of UPF1, RNA degradation can be induced in several different ways in mammals by the recruitment of the endonuclease SMG6, the SMG5-SMG7 heterodimer, and/or the decapping enhancer proline-rich nuclear receptor co-activator 2 (PNRC2) to phosphorylated UPF1. SMG6 cleaves the mRNA endonucleolytically near the terminating ribosome [32,48,49], while the SMG5/SMG7 heterodimer recruits the CCR4–NOT deadenylase complex, which catalyzes polyA tail shortening and eventually stimulates decapping of the RNA by the decapping complex [30]. A third possibility to induce RNA decay consists in the direct recruitment of the decapping complex by UPF1, either directly or indirectly in a PNRC2-dependent manner [50–53].

Finally, hydrolysis of the UPF1-bound ATP promotes the release of the NMD factors from the degrading mRNA [54], and concomitantly UPF1 is dephosphorylated by protein phosphatase 2A (PP2A) in a process that seems to require SMG5, SMG6 and SMG7 [55,56].

Thus, in a nutshell, this model proposes that correct translation termination requires the TC to be positioned in the messenger ribonucleoprotein (mRNP) spatially close to termination stimulating factors, such as for example PABPC1, whereas aberrant translation termination would be characterized by the absence of these factors and/or the presence of decay stimulating factors, such as for example UPF1 [2]. From the virus perspective, this general rule can be exploited. As discussed below, viruses have evolved a plethora of strategies to interfere with or utilize NMD to their advantage.

5. Viral mRNAs as Substrates for NMD

During evolution, the replication strategy as well as the selective pressure to maximize the coding capacity of a viral RNA (vRNA) without increasing its genome size has introduced features in vRNAs that could make it a substrate for NMD. In particular, the presence of multiple ORFs on the same RNA makes many of the TCs in vRNAs appear in positions where one would expect them to elicit NMD. Some of the viral ORFs resemble uORFs and many TCs occur in the middle of the transcript far away from a polyA tail. Figure 2 depicts a series of viral mRNAs that contain *bona fide* signatures of NMD recognition (compare Figure 2 with Figure 1). However, vRNAs flourish in the nucleus and cytoplasm of infected cells, suggesting that viruses employ mechanisms that protect their transcripts from the host cell's degradation pathways.

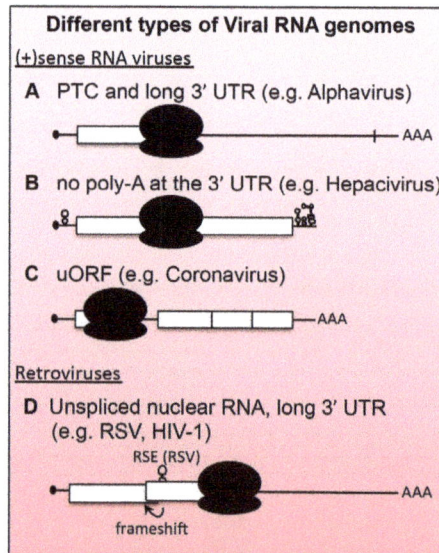

Figure 2. Different types of viral mRNAs that can be substrates for NMD. Similar to aberrant or unconventional cellular mRNA, viral transcripts contain features that make them susceptible to NMD, including the presence of PTC and long 3'-UTRs (**A**); the absence of cap-structure and polyA (**B**); uORFs (**C**); and a combination of all such as in the case of retroviruses RNAs, the translation of which also includes ribosomal frameshifts (**D**). In the case of Rous sarcoma virus (RSV), RNA secondary structures immediately downstream of the TC can act as RNA Stability Elements (RSE; **D**).

6. Documented Role of NMD and NMD Factors in Virus Infection

To date, a role for NMD or NMD factors in virus infection has been demonstrated for positive-stranded RNA ((+)RNA) viruses of animals (alphaviruses and hepatitis C virus) and plants (*Alphaflexiviridae* and *Tombusviridae*) [57,58], and for retroviruses (Rous Sarcoma virus, human T-lymphotropic virus type 1, and human immunodeficiency virus (HIV)) [59,60]. As illustrated in Figure 2 and discussed below, the mRNAs produced by each of these viruses contain features known to trigger NMD. In few cases, a viral counter-defense mechanism has been identified that protects the viral mRNA from NMD.

6.1. Positive-Stranded RNA ((+)RNA) Viruses

6.1.1. Alphaviruses and Plant (+)RNA Viruses

Recent studies have shown that NMD might constitute a conserved arm of the intrinsic innate immunity that is able to recognize and degrade vRNAs in extant eukaryotes, including mammalian cells, insects and plants [57,58,61]. Genome-wide small interfering RNA (siRNA) screens carried out in mammalian cells identified UPF1, SMG5 and SMG7 as host factors that restrict the replication of Semliki Forest virus (SFV) and Sindbis virus (SINV), two alphaviruses of the *Togaviridae* family [57].

Follow-up studies in cells infected with replication-incompetent viruses showed that depletion of UPF1 increased the half-life of the SFV genomic RNA (gRNA), indicating that this RNA is a substrate for NMD [57,62].

As for all (+)RNA viruses, the genome of alphaviruses is an mRNA-like molecule that once released into the cytoplasm of infected cells is immediately translated (Figure 3). The alphavirus genome contains two ORFs. The upstream ORF encodes the nonstructural proteins responsible for genome RNA replication and transcription. The second ORF encodes the structural proteins, capsid and envelope, that are required for the assembly of new virions (Figure 3A). Translation of the second ORF requires the synthesis of a sub-genomic mRNA encoded in the 3' third of the genome. This configuration creates a very long 3' UTR (about 4000 nucleotides) during the translation of the first ORF from the full-length gRNA (Figure 3B), which could increase susceptibility to NMD. Surprisingly however, substantial shortening of the 3' UTR failed to relieve the gRNA from the repressive effect of UPF1. Thus, it is currently not known what renders the genome of alphaviruses susceptible to NMD. In addition to the length of the 3' UTR, other factors could render the viral mRNA genome a target of NMD. Translation termination is a crucial moment for NMD activation (see Section 4). As for all other (+)RNA viruses, the replication of Alphavirus genomes requires the synthesis of a full-length complementary RNA strand (referred to as the negative-sense RNA ((-)RNA)). The viral replication complexes, encoded by the first ORF, synthesize (-)RNA in 3'-to-5' direction of the gRNA (Figure 3C). Translating ribosomes, however, move along the same RNA molecule in 5'-to-3' direction, opposite to the viral polymerase complex. This begs the question how the viral polymerase can copy the RNA genome, if the same molecule is used by ribosomes that move in opposite direction? A mechanism must exist to clear translating ribosomes from the vRNA template. The molecular details of this process are not clear. According to current models, the newly synthesized viral replicase proteins could bind to both vRNA and components of the translation machinery to trigger translation termination (or block translation re-initiation). This could create a RNP environment that triggers NMD activation. Consistent with this model, Balistreri et al. showed that impairing the helicase activity of the viral replicase complex rendered SFV hypersensitive to UPF1 depletion, which increased virus replication in primary human cells up to 20-fold [57,63]. More studies are required to shed light on this important aspect of virus replication.

In an independent study, the role of UPF1 as a restriction factor for SINV was confirmed in vivo in insects, using *Drosophila* as a model system [61]. In this study, the two ORFs of SINV were separately inserted into the cellular genome of the insect cells and the transcription of the viral RNAs was launched by an inducible promoter. In this system, inhibition of the RNA interference (RNAi) pathway provided a first means to increase virus production. In addition, expression of a dominant negative form of UPF1 increased the yields of released viruses by more than three-fold. These results are in agreement with those obtained in mammalian cells and collectively indicate that something in the genome of alphaviruses, or in the mode of translation termination, renders these RNA molecules susceptible to UPF1-mediated restriction.

In addition to alphaviruses, antiviral effects of NMD have also been shown for (+)RNA viruses of plants [58]. Starting from a genetic screen in *Arabidopsis*, this study identified UPF1, UPF3 and SMG7 as restriction factors for (+)RNA viruses of the *Alphaflexiviridae* and *Tombusviridae*. Similar to alphaviruses, these viruses also have mRNA genomes with long 3' UTRs due to the presence of multiple ORFs encoded by subgenomic RNAs. Unlike the results obtained for SFV in animal cells, however, Garcia and colleagues showed that shortening the 3' UTR of the corresponding viral transcript did increase vRNA accumulation in infected cells and resulted in more efficient virus spread. Thus, for these particular viruses, the length of the gRNA 3' UTR is an important determinant for their susceptibility to NMD. Supporting this notion, the authors showed that a (+)RNA virus of the *Potiviridae* family with a single ORF and a short 3' UTR was not restricted by the NMD.

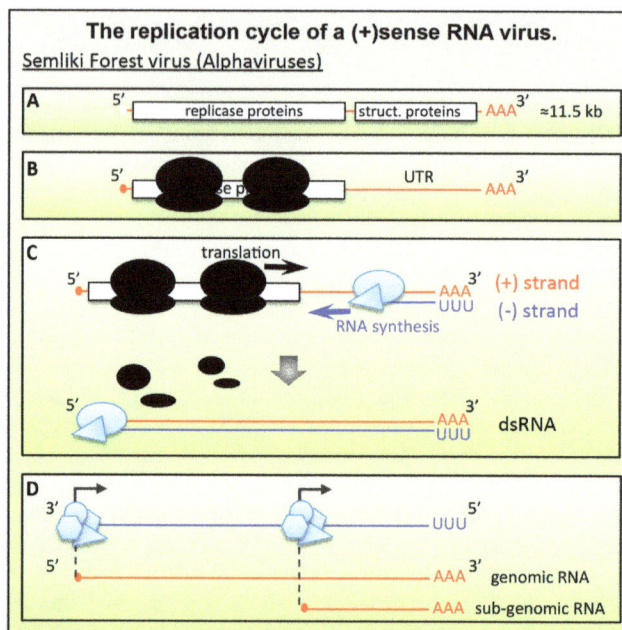

Figure 3. The replication cycle of a positive-stranded RNA ((+)RNA virus). (**A**) The bi-cystronic mRNA genome of Alphaviruses is capped a poly-adenylated. The first ORF encodes the replicase proteins; the second ORF encodes the structural genes (capsid and envelope proteins). (**B**) After virus entry, the genome is delivered into the cytoplasm where the first ORF is translated, leaving a ≈4000 nucleotides long 3'-UTR. (**C**) Newly synthesized viral replicase polyproteins assemble at the 3'-end of the genome and produce a complementary "minus" sense copy. Translation of the viral genome must be shut down. (**D**) After auto-proteolytic cleavage, the replicase complex switches the template and uses the "minus" strand to synthetize new copies of full-length genome and a shorter sub-genomic mRNA that encodes the structural proteins (the second ORF).

6.1.2. Hepatitis C Virus (HCV)

Recent studies have shown that approximately 3%–10% of the transcriptome is affected by NMD directly or indirectly in diverse eukaryotes [10]. Thus, virus induced global inhibition of the NMD pathway could lead to major rewiring of the cell, a strategy that viruses could use to create an environment favorable for virus infection. Recent evidence suggests that this could be the case for hepatitis C virus (HCV), a leading cause of liver disease. Using hepatoma cell lines, Ramage and co-authors showed that HCV infection causes a progressive inhibition of NMD activity as measured by the accumulation of three cellular RNAs known to be NMD substrates (SC35, ASNS and CARS) [64]. A combination of proteomics and RNAi-screening approaches revealed that the viral protein "core" binds to the EJC recycling factor WIBG/PYM and prevents its interaction with other components of the EJC (Y14 and Magoh). Depletions of WIBG/PYM decreased HCV replication and concomitantly suppressed the accumulation of NMD substrates. The mechanisms by which EJC components and NMD inhibition influence HCV replication remains unclear. Like other (+)RNA viruses that infect mammals, HCV is a cytoplasmic virus. Viral RNAs never reach the nucleus and do not contain introns. The inhibition of NMD observed upon HCV infection could result in the stabilization of multiple cellular mRNAs. This could contribute to creating an environment favorable to virus replication and could contribute to pathological effects associated with HCV infection [64]. Future studies are needed to shed more light into this process.

6.2. Retroviruses

6.2.1. Rous Sarcoma Virus (RSV)

Once transcribed from its integrated DNA, the full length unspliced transcript of Rous sarcoma virus (RSV) consists of a single mRNA molecule that contains several ORFs (Figure 2D). Although ribosomes translating the first ORF, *gag*, terminate about 7000 nucleotides upstream of the polyA tail and could therefore be expected to activate NMD, the unspliced full length viral RNA is surprisingly resistant to NMD [65]. However, PTCs experimentally introduced into different places of the *gag* ORF were shown to trigger NMD in chicken cells, suggesting that the natural TC of the *gag* ORF may be protected from activating NMD by specific *cis*-acting signals [65,66]. Indeed, it was found that this NMD resistance is conferred by a 400 nucleotides long sequence element located immediately downstream of *gag* TC that was termed the RNA stability element (RSE, Figure 2D) [67,68]. The RSE forms a complex RNA secondary structure [69] and includes several pyrimidine-rich stretches that were recently shown to serve as the binding platform for the polypyrimidine tract binding protein 1 (PTBP1). Recruitment of PTBP1 to the proximity of a termination codon prevents the recruitment of UPF1 and thereby antagonizes NMD, leading to the stabilization of RSV full length RNA and reporter transcripts [70]. The pyrimidine-rich sites are essential for RNA stabilization as mutations at these sites abolished protections of both viral and reporter mRNAs [70]. The authors of this study further showed that the role of PTBP1 as an NMD antagonizing factor is not limited to viral RNAs but that it can also efficiently protect cellular mRNAs from NMD when recruited downstream of an otherwise NMD-triggering TC.

6.2.2. Human T-lymphotropic Virus Type 1 (HTLV-1)

Stabilization elements such as the RSE of RSV act in *cis* to protect the mRNA molecule that harbors them. As an alternative strategy, two proteins from another retrovirus, human T-lymphotropic virus type 1 (HTLV-1), have been shown to bind to core components of the NMD machinery, causing a global inhibition of the NMD pathway. In one study, the protein TAX was shown to bind UPF1 and the translation initiation complex component INT6/eIF3E, which resulted in partial inhibition of NMD and the stabilization of viral transcripts [71]. A second study reported that the viral protein REX had similar function and was even more efficient than TAX in inhibiting NMD [72]. In this scenario, the viral proteins act in *trans* to inhibit NMD and in turn increase the half-life of viral mRNAs.

6.2.3. Human Immunodeficiency Virus Type 1 (HIV-1)

In the tug-of-war between pathogens and their host defense mechanisms, viruses often evolve means to subvert the function of cellular factors to their advantage. In the case of human immunodeficiency virus type 1 (HIV-1) for instance, two groups have shown that rather than restricting infection, UPF1 is a positive regulator of virus gene translation, vRNA nuclear export, and specific infectivity of released virions [73–75]. One of the groups reported an association and co-localization of UPF1 with the viral structural protein Gag during virus replication [73]. Depletion of UPF1 caused a reduction of the HIV-1 RNA and protein pr55Gag and UPF1 overexpression led to an up-regulation of HIV-1 expression. These UPF1-mediated effects required ongoing translation and the ATPase activity of UPF1 but not the interaction with UPF2 [73]. Interestingly, overexpression of UPF1 mutants that are inactive in NMD were still found to increase HIV-1 gene expression, suggesting a mechanism that requires the ATPase activity of the cellular enzyme but that is different from the canonical role of UPF1 in NMD. A follow-up study documented the presence of UPF1 in two distinct viral RNPs during the late replication phase [74]. One of them was detected in the cytoplasm and contained UPF1 and the viral protein Gag. The second RNP, containing UPF1, Rev, CRM+, DDX3, the nucleoporin p62, and the short isoform of UPF3a (UPF3aS), was shown to promote nucleo-cytoplasmic export of the vRNA. Interestingly, while the interaction between these RNP components and UPF1 was necessary for nuclear export of the viral genome, UPF2 and the long isoform of UPF3a (UPF3aL) were excluded from this complex. UPF2 was a negative regulator of HIV-1 RNA nuclear export. Unlike UPF3aS, UPF3L binds UPF2 and overexpression of UPF3aL also repressed the nucleus-cytoplasmic transport of the unspliced vRNA [74]. In this elegant work, the authors propose a model in which the viral protein Rev might bind UPF1 and compete for the binding of UPF2, thereby excluding this negative regulator (and its partner UPF3aL) from the viral RNP in the nucleus and promoting nuclear export of the HIV-1 RNA genome. This model implies that the overexpression of UPF2 and similar negative regulators of vRNA nuclear export could be used in conjunction with other approaches as a therapeutic strategy to interfere with HIV-1 replication [74].

Along the same lines, Serquina and colleagues confirmed that UPF1 is a positive regulator of HIV-1 infection, while the NMD-cofactor UPF2 had no role [75]. Following proteomic studies, the authors found that HIV-1 virions contain UPF1 and that viruses produced from UPF1-depleted cells are much less infectious. The loss of infectivity was not due to differences in the structure of released viruses, which contained normal ratios of genomic RNAs and correctly processed structural components. Instead, the authors found that the lack of UPF1 from virus-producing cells resulted in released virions that had impaired reverse-transcription during the following round of infection. Interestingly, the ATPase activity of UPF1 was not necessary for this step. Thus, the positive role of UPF1 on HIV-1 RNA nuclear export and translation is ATPase dependent [74], whereas the UPF1-induced increased efficiency of reverse-transcription in newly infected cells is ATPase independent [75].

In cells lacking UPF1 or expressing ATPase-defective UPF1 mutants, the infectivity of HIV-1 virions was impaired [75]. These results indicate a role of UPF1's helicase activity during virion assembly and release. The results of the two groups differ in that Serquina and colleagues did not detect an increased HIV-1 expression upon exogenous UPF1 overexpression. A possible explanation for the discrepancy between the results is that the two groups monitored different readouts and that the experiments were performed in different cells, which are known to have difference susceptibility to UPF1 depletion.

7. Conclusions and Outlook

Thus far, NMD has been viewed as a cellular process that regulates stability of certain types of aberrant mRNAs. Given the recent findings indicating an antiviral role of NMD, it has been speculated that the detection of pathogenic RNAs might have been the selective advantage driving the evolution of NMD rather than quality control of cellular mRNAs [76].

Regardless of its origin, the fact that viral genomes can be detected by NMD and that viral proteins have evolved to counteract/exploit NMD suggest that this pathway plays an important role during virus infection. Viruses will therefore become valuable tools for dissecting how NMD factors interact with and are activated by RNA substrates. Conventional tools, such as plasmid-based reporter constructs and inducible systems, provide RNA molecules that are produced in the nucleus and are therefore influenced by nuclear RNA processing and quality control mechanisms. One of the great advantages of using viruses as a tool to investigate NMD is that for many of them (e.g., (+)RNA viruses), the genome is delivered directly into the cytoplasm of the infected cells without undergoing nuclear quality control. This provides the unique opportunity to study specifically the cytoplasmic events of NMD without separated from any effects on RNA stability originating from splicing, nuclear export and RNA editing. Moreover, identification of viral factors that interfere with NMD and the characterization of the molecular mechanisms of this processes will shed new light onto different steps of the NMD mechanism, such as substrate recognition, UPF1 activation and mRNA degradation.

An important question to be addressed in the future is what triggers NMD in viral RNAs. Although many viral transcripts contain signatures known to activate NMD, a prediction cannot always be made based on the knowledge of known cellular NMD targets. In the case of alphaviruses for instance, the truncation of most of the long 3′ UTR from the viral genome did not relieve the virus from being suppressed by UPF1 [57]. This unexpected result implies that in addition to 3′ UTR length, translation termination of the viral first ORF is aberrant for another reason than the distance to the polyA tail. Future work will hopefully clarify these issues.

In addition to the NMD core components, different viruses have been recently described to specifically inhibit or degrade cellular factors involved in RNA degradation [62], which supports a general antiviral role for the cellular RNA turnover systems and indicates that in the absence of viral counter-defense strategies, viral mRNAs would become detectable and vulnerable to these quality control pathways. Understanding how cells recognize viral transcripts and how viruses avoid being detected by these degradation pathways will be of great help to further our understanding of the NMD mechanism but also possibly contribute to finding new ways of interfering with virus replication and hence prevent diseases.

Acknowledgments: We are grateful for the constructive comments of one anonymous reviewer. The research in the lab of O.M. on the link between NMD and SFV infection is supported by the NCCR RNA & Disease funded by the Swiss National Science Foundation (SNSF) and by the SNFS grant 31003A-162986. G.B. is supported by the Finnish Cancer Foundation, Finland.

Author Contributions: G.B., O.M. and C.B. wrote the manuscript.

Conflicts of Interest: The authors declare no conflict of interest.

References

1. Lykke-Andersen, J.; Bennett, E.J. Protecting the proteome: Eukaryotic cotranslational quality control pathways. *J. Cell Biol.* **2014**, *204*, 467–476. [CrossRef] [PubMed]
2. Karousis, E.D.; Nasif, S.; Muhlemann, O. Nonsense-mediated mRNA decay: novel mechanistic insights and biological impact. *Wiley interdiscip. Rev. RNA* **2016**, *7*, 661–682. [CrossRef] [PubMed]
3. He, F.; Li, X.; Spatrick, P.; Casillo, R.; Dong, S.; Jacobson, A. Genome-wide analysis of mRNAs regulated by the nonsense-mediated and 5' to 3' mRNA decay pathways in yeast. *Mol. Cell* **2003**, *12*, 1439–1452. [CrossRef]
4. Mendell, J.T.; Sharifi, N.A.; Meyers, J.L.; Martinez-Murillo, F.; Dietz, H.C. Nonsense surveillance regulates expression of diverse classes of mammalian transcripts and mutes genomic noise. *Nat.Genet.* **2004**, *36*, 1073–1078. [CrossRef] [PubMed]
5. Losson, R.; Lacroute, F. Interference of nonsense mutations with eukaryotic messenger RNA stability. *Proc. Natl. Acad. Sci. USA* **1979**, *76*, 5134–5137. [CrossRef] [PubMed]
6. Maquat, L.E.; Kinniburgh, A.J.; Rachmilewitz, E.A.; Ross, J. Unstable beta-globin mRNA in mRNA-deficient beta o thalassemia. *Cell* **1981**, *27*, 543–553. [CrossRef]
7. Culbertson, M.R.; Leeds, P.F. Looking at mRNA decay pathways through the window of molecular evolution. *Curr. Opin. Genet. Dev.* **2003**, *13*, 207–214. [CrossRef]
8. Rebbapragada, I.; Lykke-Andersen, J. Execution of nonsense-mediated mRNA decay: what defines a substrate? *Curr. Opin. Cell Biol.* **2009**, *21*, 394–402. [CrossRef] [PubMed]
9. Nicholson, P.; Yepiskoposyan, H.; Metze, S.; Zamudio Orozco, R.; Kleinschmidt, N.; Mühlemann, O. Nonsense-mediated mRNA decay in human cells: mechanistic insights, functions beyond quality control and the double-life of NMD factors. *Cell Mol. Life Sci.* **2010**, *67*, 677–700. [CrossRef] [PubMed]
10. Schweingruber, C.; Rufener, S.C.; Zund, D.; Yamashita, A.; Muhlemann, O. Nonsense-mediated mRNA decay - mechanisms of substrate mRNA recognition and degradation in mammalian cells. *Biochim. Biophys. Acta.* **2013**, *1829*, 612–623. [CrossRef] [PubMed]
11. Puretzky, A.A.; Liang, L.; Li, X.; Xiao, K.; Wang, K.; Mahjouri-Samani, M.; Basile, L.; Idrobo, J.C.; Sumpter, B.G.; Meunier, V. Low-Frequency Raman Fingerprints of Two-Dimensional Metal Dichalcogenide Layer Stacking Configurations. *ACS Nano* **2015**, *9*, 6333–6342. [CrossRef] [PubMed]
12. Lykke-Andersen, S.; Jensen, T.H. Nonsense-mediated mRNA decay: an intricate machinery that shapes transcriptomes. *Nat. Rev. Mol. Cell Bio.* **2015**, *16*, 665–677. [CrossRef] [PubMed]
13. Shaul, O. Unique Aspects of Plant Nonsense-Mediated mRNA Decay. *Trends Plant Sci.* **2015**, *20*, 767–779. [CrossRef] [PubMed]
14. He, F.; Brown, A.H.; Jacobson, A. Upf1p, Nmd2p, and Upf3p are interacting components of the yeast nonsense-mediated mRNA decay pathway. *Mol. Cell. Biol.* **1997**, *17*, 1580–1594. [CrossRef] [PubMed]

15. Min, E.E.; Roy, B.; Amrani, N.; He, F.; Jacobson, A. Yeast Upf1 CH domain interacts with Rps26 of the 40S ribosomal subunit. *RNA* **2013**, *19*, 1105–1115. [CrossRef] [PubMed]

16. He, F.; Jacobson, A. Identification of a novel component of the nonsense-mediated mRNA decay pathway by use of an interacting protein screen. *Genes Dev.* **1995**, *9*, 437–454. [CrossRef] [PubMed]

17. Clerici, M.; Deniaud, A.; Boehm, V.; Gehring, N.H.; Schaffitzel, C.; Cusack, S. Structural and functional analysis of the three MIF4G domains of nonsense-mediated decay factor UPF2. *Nucleic Acids Res.* **2014**, *42*, 2673–2686. [CrossRef] [PubMed]

18. Chakrabarti, S.; Jayachandran, U.; Bonneau, F.; Fiorini, F.; Basquin, C.; Domcke, S.; Le Hir, H.; Conti, E. Molecular mechanisms for the RNA-dependent ATPase activity of Upf1 and its regulation by Upf2. *Mol. Cell 41*, 693–703. [CrossRef] [PubMed]

19. Lykke-Andersen, J.; Shu, M.D.; Steitz, J.A. Human Upf proteins target an mRNA for nonsense-mediated decay when bound downstream of a termination codon. *Cell* **2000**, *103*, 1121–1131. [CrossRef]

20. Le Hir, H.; Sauliere, J.; Wang, Z. The exon junction complex as a node of post-transcriptional networks. *Nat. Rev. Mol. Cell Biol.* **2016**, *17*, 41–54. [CrossRef] [PubMed]

21. Kunz, J.B.; Neu-Yilik, G.; Hentze, M.W.; Kulozik, A.E.; Gehring, N.H. Functions of hUpf3a and hUpf3b in nonsense-mediated mRNA decay and translation. *RNA* **2006**, *12*, 1015–1022. [CrossRef] [PubMed]

22. Chan, W.K.; Bhalla, A.D.; Le Hir, H.; Nguyen, L.S.; Huang, L.; Gécz, J.; Wilkinson, M.F. A UPF3-mediated regulatory switch that maintains RNA surveillance. *Nat. Struct. Mol. Biol.* **2009**, *16*, 747–753. [CrossRef] [PubMed]

23. Shum, E.Y.; Jones, S.H.; Shao, A.; Dumdie, J.; Krause, M.D.; Chan, W.K.; Lou, C.H.; Espinoza, J.L.; Song, H.W.; Phan, M.H.; et al. The Antagonistic Gene Paralogs Upf3a and Upf3b Govern Nonsense-Mediated RNA Decay. *Cell* **2016**, *165*, 382–395. [CrossRef] [PubMed]

24. Denning, G.; Jamieson, L.; Maquat, L.E.; Thompson, E.A.; Fields, A.P. Cloning of a novel phosphatidylinositol kinase-related kinase: characterization of the human SMG-1 RNA surveillance protein. *J. Biol. Chem.* **2001**, *276*, 22709–22714. [CrossRef] [PubMed]

25. Hug, N.; Caceres, J.F. The RNA helicase DHX34 activates NMD by promoting a transition from the surveillance to the decay-inducing complex. *Cell Rep.* **2014**, *8*, 1845–1856. [CrossRef] [PubMed]

26. Kashima, I.; Yamashita, A.; Izumi, N.; Kataoka, N.; Morishita, R.; Hoshino, S.; Ohno, M.; Dreyfuss, G.; Ohno, S. Binding of a novel SMG-1-Upf1-eRF1-eRF3 complex (SURF) to the exon junction complex triggers Upf1 phosphorylation and nonsense-mediated mRNA decay. *Genes Dev.* **2006**, *20*, 355–367. [CrossRef] [PubMed]

27. Yamashita, A.; Izumi, N.; Kashima, I.; Ohnishi, T.; Saari, B.; Katsuhata, Y.; Muramatsu, R.; Morita, T.; Iwamatsu, A.; Hachiya, T.; et al. SMG-8 and SMG-9, two novel subunits of the SMG-1 complex, regulate remodeling of the mRNA surveillance complex during nonsense-mediated mRNA decay. *Genes Dev.* **2009**, *23*, 1091–1105. [CrossRef] [PubMed]

28. Melero, R.; Uchiyama, A.; Castaño, R.; Kataoka, N.; Kurosawa, H.; Ohno, S.; Yamashita, A.; Llorca, O. Structures of SMG1-UPFs complexes: SMG1 contributes to regulate UPF2-dependent activation of UPF1 in NMD. *Structure* **2014**, *22*, 1105–1119. [CrossRef] [PubMed]

29. Popp, M.W.; Maquat, L.E. The dharma of nonsense-mediated mRNA decay in mammalian cells. *Mol. Cells* **2014**, *37*, 1–8. [CrossRef] [PubMed]

30. Loh, B.; Jonas, S.; Izaurralde, E. The SMG5-SMG7 heterodimer directly recruits the CCR4-NOT deadenylase complex to mRNAs containing nonsense codons via interaction with POP2. *Genes Dev.* **2013**, *27*, 2125–2138. [CrossRef] [PubMed]

31. Chakrabarti, S.; Bonneau, F.; Schussler, S.; Eppinger, E.; Conti, E. Phospho-dependent and phospho-independent interactions of the helicase UPF1 with the NMD factors SMG5-SMG7 and SMG6. *Nucleic Acids Res* **2014**, *42*, 9447–9460. [CrossRef] [PubMed]

32. Eberle, A.B.; Lykke-Andersen, S.; Muhlemann, O.; Jensen, T.H. SMG6 promotes endonucleolytic cleavage of nonsense mRNA in human cells. *Nat. Struct. Mol. Biol.* **2009**, *16*, 49–55. [CrossRef] [PubMed]

33. Okada-Katsuhata, Y.; Yamashita, A.; Kutsuzawa, K.; Izumi, N.; Hirahara, F.; Ohno, S. N- and C-terminal Upf1 phosphorylations create binding platforms for SMG-6 and SMG-5:SMG-7 during NMD. *Nucleic Acids Res.* **2012**, *40*, 1251–1266. [CrossRef] [PubMed]

34. Nicholson, P.; Josi, C.; Kurosawa, H.; Yamashita, A.; Muhlemann, O. A novel phosphorylation-independent interaction between SMG6 and UPF1 is essential for human NMD. *Nucleic Acids Res.* **2014**, *42*, 9217–9235. [CrossRef] [PubMed]

35. Longman, D.; Hug, N.; Keith, M.; Anastasaki, C.; Patton, E.E.; Grimes, G.; Cáceres, J.F. DHX34 and NBAS form part of an autoregulatory NMD circuit that regulates endogenous RNA targets in human cells, zebrafish and Caenorhabditis elegans. *Nucleic Acids Res.* **2013**, *41*, 8319–8331. [CrossRef] [PubMed]

36. Anastasaki, C.; Longman, D.; Capper, A.; Patton, E.E.; Caceres, J.F. Dhx34 and Nbas function in the NMD pathway and are required for embryonic development in zebrafish. *Nucleic Acids Res.* **2011**, *39*, 3686–3694. [CrossRef] [PubMed]

37. Casadio, A.; Longman, D.; Hug, N.; Delavaine, L.; Vallejos Baier, R.; Alonso, C.R.; Cáceres, J.F. Identification and characterization of novel factors that act in the nonsense-mediated mRNA decay pathway in nematodes, flies and mammals. *EMBO Rep.* **2015**, *16*, 71–78. [CrossRef] [PubMed]

38. Nakamura, Y.; Ito, K.; Matsumura, K.; Kawazu, Y.; Ebihara, K. Regulation of translation termination: conserved structural motifs in bacterial and eukaryotic polypeptide release factors. *Biochem. Cell Biol.* **1995**, *73*, 1113–1122. [CrossRef] [PubMed]

39. Pisarev, A.V.; Skabkin, M.A.; Pisareva, V.P.; Skabkina, O.V.; Rakotondrafara, A.M.; Hentze, M.W.; Hellen, C.U.; Pestova, T.V. The role of ABCE1 in eukaryotic posttermination ribosomal recycling. *Mol. Cell* **2010**, *37*, 196–210. [CrossRef] [PubMed]

40. Barthelme, D.; Dinkelaker, S.; Albers, S.V.; Londei, P.; Ermler, U.; Tampé, R. Ribosome recycling depends on a mechanistic link between the FeS cluster domain and a conformational switch of the twin-ATPase ABCE1. *Proc. Natl. Acad. Sci. USA* **2011**, *108*, 3228–3233. [CrossRef] [PubMed]

41. Ivanov, A.; Mikhailova, T.; Eliseev, B.; Yeramala, L.; Sokolova, E.; Susorov, D.; Shuvalov, A.; Schaffitzel, C.; Alkalaeva, E. PABP enhances release factor recruitment and stop codon recognition during translation termination. *Nucleic Acids Research* **2016**, *44*, 7766–7776. [CrossRef] [PubMed]

42. Singh, G.; Rebbapragada, I.; Lykke-Andersen, J. A competition between stimulators and antagonists of Upf complex recruitment governs human nonsense-mediated mRNA decay. *PLoS Biol.* **2008**, *6*, e111. [CrossRef] [PubMed]

43. Silva, A.L.; Ribeiro, P.; Inacio, A.; Liebhaber, S.A.; Romao, L. Proximity of the poly(A)-binding protein to a premature termination codon inhibits mammalian nonsense-mediated mRNA decay. *RNA* **2008**, *14*, 563–576. [CrossRef] [PubMed]

44. Muhlemann, O.; Jensen, T.H. mRNP quality control goes regulatory. *Trends Genet.* **2012**, *28*, 70–77. [CrossRef] [PubMed]

45. Maquat, L.E. Nonsense-mediated mRNA decay: splicing, translation and mRNP dynamics. *Nat. Rev. Mol. Cell Biol.* **2004**, *5*, 89–99. [CrossRef] [PubMed]

46. Amrani, N.; Ganesan, R.; Kervestin, S.; Mangus, D.A.; Ghosh, S.; Jacobson, A. A faux 3'-UTR promotes aberrant termination and triggers nonsense-mediated mRNA decay. *Nature* **2004**, *432*, 112–118. [CrossRef] [PubMed]

47. Peixeiro, I.; Silva, A.L.; Romao, L. Control of human beta-globin mRNA stability and its impact on beta-thalassemia phenotype. *Haematologica* **2011**, *96*, 905–913. [CrossRef] [PubMed]

48. Lykke-Andersen, S.; Chen, Y.; Ardal, B.R.; Lilje, B.; Waage, J.; Sandelin, A.; Jensen, T.H. Human nonsense-mediated RNA decay initiates widely by endonucleolysis and targets snoRNA host genes. *Genes Dev.* **2014**, *28*, 2498–2517. [CrossRef] [PubMed]

49. Boehm, V.; Haberman, N.; Ottens, F.; Ule, J.; Gehring, N.H. 3' UTR length and messenger ribonucleoprotein composition determine endocleavage efficiencies at termination codons. *Cell Rep.* **2014**, *9*, 555–568. [CrossRef] [PubMed]

50. Lykke-Andersen, J. Identification of a human decapping complex associated with hUpf proteins in nonsense-mediated decay. *Mol. Cell. Biol.* **2002**, *22*, 8114–8121. [CrossRef] [PubMed]

51. Cho, H.; Kim, K.M.; Kim, Y.K. Human proline-rich nuclear receptor coregulatory protein 2 mediates an interaction between mRNA surveillance machinery and decapping complex. *Mol. Cell* **2009**, *33*, 75–86. [CrossRef] [PubMed]

52. Lai, T.; Cho, H.; Liu, Z.; Bowler, M.W.; Piao, S.; Parker, R.; Kim, Y.K.; Song, H. Structural basis of the PNRC2-mediated link between mrna surveillance and decapping. *Structure* **2012**, *20*, 2025–2037. [CrossRef] [PubMed]

53. Cho, H.; Han, S.; Choe, J.; Park, S.G.; Choi, S.S.; Kim, Y.K. SMG5-PNRC2 is functionally dominant compared with SMG5-SMG7 in mammalian nonsense-mediated mRNA decay. *Nucleic Acids Res.* **2013**, *41*, 1319–1328. [CrossRef] [PubMed]

54. Franks, T. M.; Singh, G.; Lykke-Andersen, J. Upf1 ATPase-dependent mRNP disassembly is required for completion of nonsense- mediated mRNA decay. *Cell* **2010**, *143*, 938–950. [CrossRef] [PubMed]

55. Ohnishi, T.; Yamashita, A.; Kashima, I.; Schell, T.; Anders, K.R.; Grimson, A.; Hachiya, T.; Hentze, M.W.; Anderson, P.; Ohno, S. Phosphorylation of hUPF1 induces formation of mRNA surveillance complexes containing hSMG-5 and hSMG-7. *Mol. Cell* **2003**, *12*, 1187–1200. [CrossRef]

56. Chiu, S.Y.; Serin, G.; Ohara, O.; Maquat, L.E. Characterization of human Smg5/7a: a protein with similarities to Caenorhabditis elegans SMG5 and SMG7 that functions in the dephosphorylation of Upf1. *RNA* **2003**, *9*, 77–87. [CrossRef] [PubMed]

57. Balistreri, G.; Horvath, P.; Schweingruber, C.; Zünd, D.; McInerney, G.; Merits, A.; Mühlemann, O.; Azzalin, C.; Helenius, A. The host nonsense-mediated mRNA decay pathway restricts Mammalian RNA virus replication. *Cell Host Microbe* **2014**, *16*, 403–411. [CrossRef] [PubMed]

58. Garcia, D.; Garcia, S.; Voinnet, O. Nonsense-mediated decay serves as a general viral restriction mechanism in plants. *Cell Host Microbe* **2014**, *16*, 391–402. [CrossRef] [PubMed]

59. Quek, B.L.; Beemon, K. Retroviral strategy to stabilize viral RNA. *Curr. Opin. Microbiol.* **2014**, *18*, 78–82. [CrossRef] [PubMed]

60. Mocquet, V.; Durand, S.; Jalinot, P. How Retroviruses Escape the Nonsense-Mediated mRNA Decay. *AIDS Res. Hum. Retroviruses* **2015**, *31*, 948–958. [CrossRef] [PubMed]

61. Wernet, M.F.; Klovstad, M.; Clandinin, T.R. Generation of infectious virus particles from inducible transgenic genomes. *Curr. Biol.* **2014**, *24*, R107–R108. [CrossRef] [PubMed]

62. Moon, S.L.; Wilusz, J. Cytoplasmic viruses: rage against the (cellular RNA decay) machine. *PLoS Pathog.* **2013**, *9*, e1003762. [CrossRef] [PubMed]

63. Balistreri, G.; Caldentey, J.; Kaariainen, L.; Ahola, T. Enzymatic defects of the nsP2 proteins of Semliki Forest virus temperature-sensitive mutants. *J. Virol* **2007**, *81*, 2849–2860. [CrossRef] [PubMed]

64. Ramage, H.R.; Kumar, G.R.; Verschueren, E.; Johnson, J.R.; Von Dollen, J.; Johnson, T.; Newton, B.; Shah, P.; Horner, J.; Krogan, N.J.; et al. A combined proteomics/genomics approach links hepatitis C virus infection with nonsense-mediated mRNA decay. *Mol. Cell* **2015**, *57*, 329–340. [CrossRef] [PubMed]

65. Barker, G.F.; Beemon, K. Rous sarcoma virus RNA stability requires an open reading frame in the gag gene and sequences downstream of the gag-pol junction. *Mol. Cell. Biol.* **1994**, *14*, 1986–1996. [CrossRef] [PubMed]

66. LeBlanc, J.J.; Beemon, K.L. Unspliced Rous sarcoma virus genomic RNAs are translated and subjected to nonsense-mediated mRNA decay before packaging. *J. Virol* **2004**, *78*, 5139–5146. [CrossRef] [PubMed]

67. Weil, J.E.; Beemon, K.L. A 3' UTR sequence stabilizes termination codons in the unspliced RNA of Rous sarcoma virus. *RNA* **2006**, *12*, 102–110. [CrossRef] [PubMed]

68. Withers, J.B.; Beemon, K.L. Structural features in the Rous sarcoma virus RNA stability element are necessary for sensing the correct termination codon. *Retrovirology* **2012**, *7*, 65. [CrossRef] [PubMed]

69. Withers, J.B.; Beemon, K.L. The structure and function of the rous sarcoma virus RNA stability element. *J. Cell Biochem.* **2011**, *112*, 3085–3092. [CrossRef] [PubMed]

70. Ge, Z.; Quek, B.L.; Beemon, K.L.; Hogg, J.R. Polypyrimidine tract binding protein 1 protects mRNAs from recognition by the nonsense-mediated mRNA decay pathway. *Elife* **2016**, *5*. [CrossRef] [PubMed]

71. Mocquet, V.; Neusiedler, J.; Rende, F.; Cluet, D.; Robin, J.P.; Terme, J.M.; Duc Dodon, M.; Wittmann, J.; Morris, C.; Le Hir, H.; et al. The human T-lymphotropic virus type 1 tax protein inhibits nonsense-mediated mRNA decay by interacting with INT6/EIF3E and UPF1. *J Virol* **2012**, *86*, 7530–7543. [CrossRef] [PubMed]

72. Nakano, K.; Ando, T.; Yamagishi, M.; Yokoyama, K.; Ishida, T.; Ohsugi, T.; Tanaka, Y.; Brighty, D.W.; Watanabe, T. Viral interference with host mRNA surveillance, the nonsense-mediated mRNA decay (NMD) pathway, through a new function of HTLV-1 Rex: implications for retroviral replication. *Microbes Infect.* **2013**, *15*, 491–505. [CrossRef] [PubMed]

73. Ajamian, L.; Abrahamyan, L.; Milev, M.; Ivanov, P.V.; Kulozik, A.E.; Gehring, N.H.; Mouland, A.J. Unexpected roles for UPF1 in HIV-1 RNA metabolism and translation. *RNA* **2008**, *14*, 914–927. [CrossRef] [PubMed]

74. Ajamian, L.; Abel, K.; Rao, S.; Vyboh, K.; García-de-Gracia, F.; Soto-Rifo, R.; Kulozik, A.E.; Gehring, N.H.; Mouland, A.J. HIV-1 Recruits UPF1 but Excludes UPF2 to Promote Nucleocytoplasmic Export of the Genomic RNA. *Biomolecules* **2015**, *5*, 2808–2839. [CrossRef] [PubMed]

75. Serquina, A.K.; Das, S.R.; Popova, E.; Ojelabi, O.A.; Roy, C.K.; Göttlinger, H.G. UPF1 is crucial for the infectivity of human immunodeficiency virus type 1 progeny virions. *J. Virol* **2013**, *87*, 8853–8861. [CrossRef] [PubMed]
76. Hamid, F.M.; Makeyev, E.V. Exaptive origins of regulated mRNA decay in eukaryotes. *Bioessays* **2016**, *38*, 830–838. [CrossRef] [PubMed]

viruses

MDPI

Review

Attacked from All Sides: RNA Decay in Antiviral Defense

Jerome M. Molleston and Sara Cherry *

Department of Microbiology, University of Pennsylvania Perelman School of Medicine,
Philadelphia, PA 19104, USA; jeromem@mail.med.upenn.edu
* Correspondence: cherrys@mail.med.upenn.edu; Tel.: +1-215-746-2384

Academic Editors: Karen Beemon and Robert Hogg
Received: 9 November 2016; Accepted: 26 December 2016; Published: 4 January 2017

Abstract: The innate immune system has evolved a number of sensors that recognize viral RNA (vRNA) to restrict infection, yet the full spectrum of host-encoded RNA binding proteins that target these foreign RNAs is still unknown. The RNA decay machinery, which uses exonucleases to degrade aberrant RNAs largely from the 5′ or 3′ end, is increasingly recognized as playing an important role in antiviral defense. The 5′ degradation pathway can directly target viral messenger RNA (mRNA) for degradation, as well as indirectly attenuate replication by limiting specific pools of endogenous RNAs. The 3′ degradation machinery (RNA exosome) is emerging as a downstream effector of a diverse array of vRNA sensors. This review discusses our current understanding of the roles of the RNA decay machinery in controlling viral infection.

Keywords: RNA decay; RNA-protein interactions; decapping; Xrn1; exosome; TRAMP; exonuclease; RNAse; intrinsic immunity; antiviral

---

1. Viral RNAs (vRNAs) as Foreign RNAs

RNA viruses produce RNAs which differ substantially from normal cellular RNAs, leading to their recognition by host-encoded RNA-binding proteins. Unlike cellular RNAs, the genomes of RNA viruses are replicated by RNA-dependent RNA polymerases (RdRp) through antigenome intermediates, creating both transient double-stranded RNA (dsRNA) structures and 5′ triphosphate ends not normally present in cellular messenger RNAs (mRNAs) [1]. For 5′ end protection and recruitment of the translational machinery, endogenous mRNAs are capped in the nucleus. However, cytoplasmic RNA viruses have no access to the normal cellular capping machinery and, thus, many of these viruses go to great lengths to acquire a cap, including enzymatically synthesizing a cap or cap mimic, or acquiring one from cellular mRNA through a process known as cap-snatching [2]. For translation and 3′ end protection, most mRNAs are polyadenylated; some cytoplasmic viral mRNAs achieve this through RdRp-mediated polyadenylation [3]. Furthermore, some viruses encode 3′ structures which impede exonucleases [4,5]. These are but a few examples of the complexity of viral RNA (vRNA) metabolism, which can render viruses susceptible to both immune sensors and the cellular RNA decay machinery.

2. Innate Immune Recognition of vRNA

2.1. RIG-I Like Receptors (RLRs) and DEAD-Box Helicases

The RNAs produced during viral replication serve as an important sign of infection, and a series of sensors have evolved to detect these RNAs. In mammals, many cytosolic vRNAs are recognized by the RIG-I-like receptors (RLRs), RIG-I and MDA-5, which are homologous DEAD-box RNA helicases [6]. Each recognizes different RNA structures. RIG-I recognizes short dsRNAs and RNAs with 5′ triphosphates, and plays a role in restricting viruses, including paramyxoviruses, orthomyxoviruses, and flaviviruses [7,8]. In contrast, MDA-5 recognizes longer dsRNAs and higher-order RNA structures, and is integral for recognition of picornaviruses [9,10]. Both can respond to the synthetic dsRNA polyinosinic-polycytidylic acid (poly(I:C)) dependent on length; long poly(I:C) is a ligand of MDA-5, while short poly(I:C) can activate RIG-I [11]. The RLRs primarily act by signaling the interferon (IFN) system through their adaptor, MAVS; IFN then signals an extensive transcriptional antiviral program [12–15]. A third RLR helicase, LGP2, lacks the ability to signal through MAVS, and in different contexts has been found to either inhibit or potentiate the signaling of the other two RLRs [16–22]. In response, many RNA viruses have evolved mechanisms to evade the IFN system and, thus, avoid the consequences of RLR detection; for example, both Rift Valley fever virus (RVFV) and Sindbis virus (SINV) encode accessory proteins (NSs and nsP2) which inhibit the transcription of IFNs [23–27].

Just as mammalian cells utilize the RLR DEAD-box RNA helicases to recognize cytosolic vRNA, the closest *Drosophila* homolog, Dicer-2, recognizes viral dsRNA intermediates generated during infection [28–30]. However, Dicer-2 functions as both sensor and effector; in addition to its helicase domain, it has a ribonuclease (RNase) III domain which cleaves dsRNAs into siRNAs which, in turn, are loaded into the Argonaute 2-containing RNA-induced silencing complex, preventing RNA translation and cleaving vRNA [31]. The antiviral RNA silencing pathway in *Drosophila* is essential for immune defense; flies with mutations in this pathway rapidly succumb to viral infection. To evade this immune defense, natural insect pathogens such as *Drosophila* C virus encode suppressors of RNAi [29,32]. Dicer-2 also has silencing-independent antiviral functions which closely parallel the signaling functions of the RLRs; Dicer-2 is required to induce the transcription of the antiviral factor Vago, suggesting that it is also a regulator of antiviral transcription during viral infection in insects [33].

New roles are continually emerging for the larger family of DEAD-box helicases in recognizing vRNA. Many of these genes have roles both in normal cellular metabolism as well as in the control of viral infection. For example, DDX17 normally binds stem-loop structures of primary microRNAs (pri-miRNAs) in the nucleus and recruits the Drosha-anchored microprocessor complex to produce pre-miRNAs [34–37]. However, upon viral infection, DDX17 is repurposed and exported to the cytoplasm, where it binds a stem-loop miRNA-like structure in RVFV RNA in order to restrict viral replication [37]. Additional helicases are involved in innate recognition, such as the DEAD-box helicase DDX60, which interacts with RIG-I, and the complex of DDX1, DDX21, and DHX36, which bind the innate immune adaptor TRIF, in each case facilitating their activity [38,39].

2.2. Toll-Like Receptors (TLRs)

Endosomal RNAs are sensed by toll-like receptors (TLRs) including TLR3 and TLR7, which detect dsRNA and ssRNA, respectively. These sensors signal through the adaptors TRIF and MyD88 to activate antiviral transcription programs [6]. Endosomal TLRs are highly expressed in dedicated immune cells such as dendritic cells, but are missing from many other cell types, which can only effectively sense cytosolic RNA [40–42].

2.3. Protein Kinase R (PKR)

Mammalian cells possess additional cytosolic sensors of vRNA including the dsRNA-activated protein kinase R (PKR) [43,44]. Activation of PKR by dsRNAs from viruses or poly(I:C) induces autophosphorylation of PKR and subsequent phosphorylation of eIF2α, shutting down global protein translation thereby preventing viral protein synthesis [45–48]. Many viruses prevent PKR-mediated translational shutdown by binding dsRNA or PKR to prevent its activation [44,49]. Other viruses, such as RVFV and poliovirus, induce PKR degradation [50–52]. In contrast, several other viruses, such as hepatitis C virus (HCV) and SINV, encode RNA structures which bypass the PKR-dependent global translational arrest and continue to be efficiently translated under these stress conditions [27,53–56]. Moreover, studies have found that RLR-dependent transcriptional activation is dependent on PKR and vice-versa, suggesting crosstalk between these pathways [57,58].

2.4. Ribonucleases (RNases)

Another cytoplasmic sensor of viral dsRNA, $2'$-$5'$ oligoadenylate synthetase (OAS), is an IFN-inducible enzyme and, thus, is up-regulated in response to viral detection by sensors, such as the RLRs and TLRs [59,60]. Upon sensing vRNA, OAS synthesizes 2,5-adenylate, which, in turn, activates the latent cytoplasmic endoribonuclease RNASEL. RNASEL cleaves vRNA and cellular RNA, thereby inhibiting viral replication [61–63]. Furthermore, these cleaved RNAs can, in turn, act as substrates for RLR detection, amplifying the antiviral program [64]. In addition, RNASEL promotes apoptosis in response to viral infection, preventing further viral spread [65]. Recent studies have begun to characterize RNAs as more or less susceptible to RNASEL and to postulate further functions for RNASEL-mediated regulation of specific RNAs [66].

Drosha, a nuclear RNase III enzyme, has a canonical role in processing pri-miRNAs to pre-miRNAs before they are exported to the cytoplasm for Dicer processing. Recent studies have shown that Drosha has antiviral activity [67]. Drosha is exported to the cytoplasm in response to diverse RNA viruses, and restricts RNA virus infection by unknown mechanisms, although it is likely that Drosha is recognizing stem-loop structures in vRNA.

New research continues to uncover RNase activity among previously identified antiviral proteins. SAMHD1 was identified as a restriction factor for human immunodeficiency virus (HIV) that is antagonized by the viral protein Vpx [68]. An initial search for the antiviral mechanism revealed that SAMHD1 is a deoxynucleotide triphosphohydrolase that degrades DNA nucleotides, restricting the nucleotide pool available to HIV [69,70]. However, recent work has also identified $3'$-$5'$ DNase and RNase activity for SAMHD1, suggesting additional antiviral functions that may be active against HIV and other viruses [71].

3. The Canonical RNA Decay Machinery and vRNA Targeting

Emerging data suggest that the canonical RNA decay machinery, which is largely dependent on exonucleases, also plays an important role in antiviral immunity. In general, RNA decay proceeds from either the $5'$ or $3'$ end of an RNA transcript, and has roles in RNA biogenesis, RNA quality control, and normal RNA turnover. Ribosomal RNAs (rRNAs), small nuclear RNAs (snRNAs), and small nucleolar RNAs (snoRNAs) all require RNA processing in the nucleus to reach their mature forms [72]. Quality control also begins in the nucleus, where RNAs which fail to be properly matured, such as hypoadenylated mRNAs and hypomodified tRNAs, are degraded before they can leave the nucleus [73,74]. In the cytoplasm, additional quality-control checkpoints, such as nonsense-mediated decay or no-go decay, detect stalled ribosomes or premature stop codons and degrade these aberrant mRNAs to release and recycle the translational machinery [75]. Recent studies have shown that RNA decay machinery also serves a key role in post-transcriptional regulation of groups of RNAs, called regulons, which are rapidly co-regulated through specific recognition of sequences in their $5'$ and $3'$ untranslated regions (UTRs) [76]. These include sequences such as terminal oligopyrimidines (TOPs) at the $5'$ and AU-rich elements (AREs) at the $3'$ ends of RNAs [77–79]. As viruses possess many features of aberrant RNAs, they are increasingly recognized as targets of the RNA decay machinery.

4. Antiviral Roles for Nonsense-Mediated Decay (NMD)

Nonsense-mediated decay (NMD) is the process by which mRNAs with stop codons far from the $3'$ end of an mRNA, either due to a premature stop codon or a long $3'$ UTR, are detected and degraded at either the $5'$ or $3'$ ends [75,80,81]. Several RNA viruses have been shown to be sensitive to this pathway. In particular, the NMD components Upf1, Smg5, and Smg7 restrict the replication of Semliki Forest virus (SFV) in mammalian cells [82]. The mechanism of this restriction is unclear, and may act through degradation of vRNA or indirect effects; it is independent of viral $3'$ UTR length, suggesting that the long $3'$ UTR of SFV mRNA is not necessary for NMD sensitivity. The antiviral effect of NMD is ancient; several plant NMD orthologs, including Upf1, were found to restrict potato virus X by recognizing vRNAs with long 3′ UTRs [83]. NMD is also antagonized by viruses; both HCV and human T-cell leukemia virus type 1 produce proteins which inhibit NMD, suggesting evolutionary pressure to evade this antiviral mechanism [84,85].

5. 5′ Decapping and Decay

Endogenous mRNAs targeted for 5′ decay are typically first deadenylated by the CCR4-NOT complex, often assisted by other deadenylating enzymes, before they can be targeted for decay [80,86]. Although deadenylation is the first regulated step towards mRNA degradation, it is sometimes reversible, and can act to "pause" mRNA translation rather than degrading these targets [87]. Removal of the 5′ cap of RNA by decappers (e.g., Dcp2) is irreversible, and permits degradation by the 5′–3′ exonucleases Xrn1 and Xrn2, in the cytoplasm and nucleus, respectively [88–90]. This process is largely conserved from yeast to mammals, though mammals have evolved multiple, partially-redundant decapping enzymes with specificity for different targets; Dcp2 is preferentially utilized in NMD and Nudt16 is preferentially involved in degradation of mRNAs containing AREs or 5′ TOPs [77,91].

Deadenylation, decapping, and 5′ degradation activities coalesce in ribonucleoprotein (RNP) structures known as processing bodies (P bodies). These RNPs consist of mRNAs targeted for decay as well as components of the decapping and 5′ degradation machinery (including Dcp2, its activators, and Xrn1) [86,92]. RNAs that accumulate in P bodies are removed from translation, and normally degraded [93]. Although P-bodies are present in normal cells at baseline, their number and size increase in response to a variety of stressors [94,95]. P bodies can interact with and exchange RNAs with other RNP granules, such as stress granules, which are composed of translationally-stalled RNAs and chaperone proteins [95,96]. There is evidence that visible P bodies are a consequence of high concentrations of mRNAs undergoing decay rather than being necessary for decay, as P body structure can be disrupted without preventing RNA degradation [97,98]. Furthermore, up-regulation of 5′ decapping and decay leads to the loss of visible P bodies due to the depletion of RNA targets [77]. These data suggest that P bodies are dynamic structures which form and dissolve in response to RNA target levels.

6. Antiviral Roles for 5′ to 3′ RNA Decay

The 5′ to 3′ RNA decay machinery can inhibit viral replication in a number of different ways (Figure 1). Studies have shown that the cytoplasmic 5′ RNA exonuclease, Xrn1, can target flavivirus RNAs and in response these viruses antagonize Xrn1 by encoding structured RNAs that result in Xrn1 stalling [99–101]. Furthermore, poliovirus induces the degradation of host 5′ decay factors, such as Xrn1 and Dcp2, through a combination of viral and host proteases, suggesting evolutionary pressure to evade host 5′ RNA decay machinery [102]. In addition to directly targeting vRNAs, the 5′ decay machinery can also impact viral replication indirectly. Recent studies have shown that decappers limit the pool of host mRNAs available for RVFV to cap snatch from, attenuating replication in both insects and mammals [77,103]. Additionally, in mammals, RVFV infection induces NUDT16-mediated decapping and decay of 5′ TOP-containing mRNAs encoding the translational machinery, limiting both global and virus-specific translation [77].

P body structure is also altered during many viral infections. The up-regulation of 5′ decay during RVFV infection prevents the formation of P bodies due to depletion of the RNA targets around which they nucleate [77]. Poliovirus induces the degradation of 5′ decay proteins such as Xrn1 and Dcp1a, preventing P body formation [102]. In addition to the destruction of P bodies during some viral infections, P body components can be repurposed by viruses to facilitate infection. Flaviviruses relocalize P body components to viral replication centers, where they bind viral 3′ UTRs, promoting efficient viral replication [104–106].

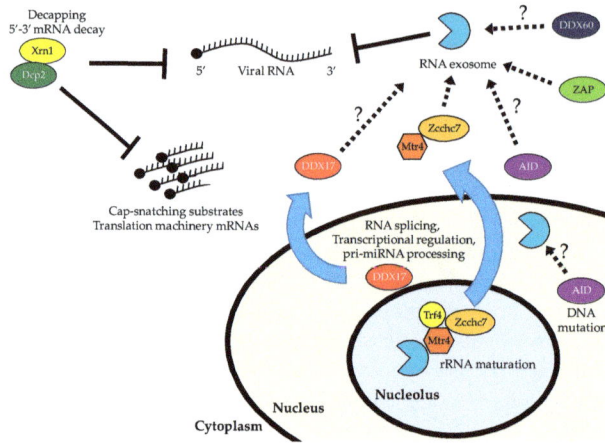

Figure 1. The 5′ to 3′ decay machinery can inhibit viral infection directly through degradation of viral RNA (vRNA; flaviviruses) or indirectly through decapping and degradation of RNAs needed for viral transcription and translation (bunyaviruses). The 3′ to 5′ decay machinery, the RNA exosome, interacts with a variety of RNA-binding proteins, some of which are exported to the cytoplasm in response to viral infection. Recruitment of the exosome can result in degradation of vRNA.

7. The RNA Exosome

3′ to 5′ degradation is largely mediated by the RNA exosome. This complex consists of a hexameric barrel (six proteins with RNase PH homology) and a cap structure (three proteins with S1 RNA-binding domains) [107–109]. These structural components of the exosome form an internal channel wide enough to permit entry of single-stranded but not double-stranded RNA [110]. In contrast to the 5′ decay machinery which localizes to cytoplasmic P bodies, exosome components are both nuclear and cytoplasmic, and can accumulate in poorly-understood RNP granules [92,111–113]. While exosome proteins share structural and sequence homology to RNases, the structural components of the exosome are not believed to contribute directly to RNA degradation in vivo. Rather, in yeast, where it has been extensively characterized, 3′ to 5′ exonuclease activity is performed by two exosome-associated RNA exonucleases: Rrp6, which is exclusive to the nucleus, and Dis3, which is present in both the nucleus and the cytoplasm [114–116].

The exosome is both structurally and functionally conserved from yeast to humans [117]. Indeed, mutations in the yeast exosomal genes Rrp4, Csl4, or Dis3 can be complemented with the human orthologs [116]. Interestingly, the localization of the exosome exonucleases has diverged over evolutionary history. Rrp6 is present in both the nuclear and cytoplasmic fractions of human cells, and Dis3 has two additional paralogs in humans, Dis3L1 and Dis3L2, which function exclusively in the cytoplasm [118]. Dis3L2, which lacks the exosome-associating PIN domain, operates independently of the larger exosome complex in a separate 3′ to 5′ degradation system which favors terminally uridylated RNAs [119–121].

Though the exosome degrades RNAs indiscriminately in vitro, it degrades RNAs in vivo in a regulated fashion by relying on RNA-binding cofactor complexes that bind specific targets and recruit the exosome for degradation [122]. All known exosome cofactor complexes are anchored by helicases which are thought to unwind higher-order RNA structures to permit single-stranded RNA to be inserted into the exosome barrel for decay [123]. Different RNAs are targeted by the exosome in the nucleus, nucleolus, and cytoplasm; therefore, the exosome relies on different cofactors in each subcellular compartment to target these diverse RNAs for decay. Two major complexes, the cytoplasmic Ski and nuclear TRAMP complexes, have been extensively characterized in yeast.

The Superkiller (Ski) complex is the major cytoplasmic exosome cofactor complex in yeast, named for the "superkilling" phenotype of dsRNA viruses, which are lethal to yeast deficient in these genes [124]. The Ski genes were identified before the discovery of the exosome, and some components of the exosome barrel were also assigned Ski names. Though mutants in cofactor Ski genes lead to increased vRNA, this has not yet been definitively linked to exosomal RNA degradation [125]. The Ski complex consists of a DExH/D-box helicase, Ski2, a tetratricopeptide repeat-containing protein, Ski3, and a WD repeat-containing protein, Ski8 [126]. An adaptor G-protein, Ski7, physically links the Ski complex and the exosome and is required for Ski complex-mediated decay [127]. The Ski complex is involved in recruiting the exosome to RNAs targeted for NMD, as well as nonstop decay [128–130]. Orthologs for all three Ski genes are present in higher organisms, though their specific targets have not been clearly defined [131,132]. Interestingly, a recent paper found that the human Ski complex-associated helicase SKI2L prevents hyper-activation of RIG-I in uninfected cells, protecting cells from autoimmune activation and patients harboring mutations in this gene presented with anomalously high IFN signatures [133]. Though the exosome was not shown to be required for this activity, it does suggest that the Ski-associated helicase, potentially with the exosome, may serve to protect the intracellular milieu from overactive RIG-I signaling, paralleling the role of the DNA exonuclease TREX, which degrades cytoplasmic DNA to prevent hyper-activation of the DNA sensor cGAS [134–136].

The yeast TRAMP (Trf4/5-Air1/2-Mtr4-Polyadenylation) complex, located in the nucleus, has known roles in degrading hypomodified tRNAs, hypoadenylated mRNAs, cryptic unstable transcripts (CUTs), and in the biogenesis of rRNA, snRNA, and snoRNA [72–74,122,137,138]. The complex is anchored by a DExD/H box helicase, Mtr4, which binds the other TRAMP components through its arch domain [123,139]. The Zn-knuckle RNA-binding proteins Air1 and Air2 bind specific RNAs and target them for degradation [140]. These two proteins are partially functionally redundant; mutants in each protein accumulate overlapping but non-identical populations of snRNAs, snoRNAs, and mRNAs, and double-mutant strains fail to grow. Trf4 and Trf5 are non-canonical poly(A) polymerases which add 5–6 adenines to RNAs bound to the TRAMP complex [137]. The addition of short poly(A) tails creates an unstructured 3' end which is thought to facilitate insertion of the RNA into the exosome barrel [141]. This adenylation parallels the role of polyadenylation in *Escherichia coli*, which, unlike eukaryotic polyadenylation, targets RNAs for decay [142,143]. In addition to the canonical TRAMP complex, Mtr4 can form other modular cofactor complexes by associating with the adaptors Nop53 or Utp18, which assist in some rRNA maturation steps [144].

The TRAMP complex is conserved in humans, but nuclear RNA degradation machinery has additional complexity. As in yeast, human TRAMP is composed of a helicase, hMTR4, a zinc-finger Air-like protein, hZCCHC7, and a poly(A) polymerase, hTRF4-1 or hTRF4-2 [145,146]. However, unlike yeast TRAMP, the human TRAMP complex is restricted to the nucleolus, and is only known to process rRNA [145,147]. Most of the yeast TRAMP targets, such as mRNAs, snRNAs, snoRNAs, and promoter upstream transcripts (PROMPTs, which are analogous to yeast CUTs) appear to be regulated in human cells by the nuclear exosome targeting (NEXT) complex, which shares hMTR4 with the TRAMP complex, but also contains the zinc-finger protein hZCCHC8 and the RNA-binding motif protein hRBM7 [145, 148,149]. Other targets are likely to exist for mammalian TRAMP-like complexes; murine cells depleted of Mtr4 accumulate adenylated 5' miRNA fragments, suggesting that adenylation and Mtr4-mediated degradation may be important for these RNAs [150]. The full spectrum of Mtr4-anchored complexes in mammals and the regulation of other classical yeast TRAMP targets (such as misprocessed tRNAs) remain unclear.

8. Antiviral Roles for the RNA Exosome and 3' Decay

Studies have implicated the exosome in antiviral defense. A number of antiviral RNA-binding proteins co-immunoprecipitate with the exosome, suggesting that their mechanism of action may involve exosomal degradation. DDX17 restricts RVFV by binding a miRNA-like stem loop structure encoded in the vRNA [37]. Though its mechanism of restriction is unknown, DDX17 binds to the exosome, suggesting that it directly recruits the exosome to degrade these bound vRNAs [145,151]. DDX60, which is antiviral against vesicular stomatitis virus (VSV), also binds the exosome [39]. However, DDX60 does not depend on the exosome for its antiviral function, but rather bridges vRNA and RLRs to potentiate signaling. The cytidine deaminase AID, which binds the exosome and hepatitis B virus (HBV) RNA in a complex, is antiviral when overexpressed only if the exosome is present, suggesting the possibility that it recruits the exosome to degrade HBV RNA [152]. The zinc-finger antiviral protein (ZAP) binds SINV and retrovirus RNA, as well as components of the exosome [153]. In overexpression systems, ZAP restricts MLV viral replication in an exosome-dependent fashion, as well as affecting the expression and stability of viral luciferase reporters for both MLV and HIV [154,155]. It remains unclear if the exosome is required for the activity of endogenous ZAP or degrades ZAP-bound vRNAs. The cell biology of these factors is largely unexplored, but overexpressed DDX60 and ZAP localize to the cytoplasm, while overexpressed AID binds the exosome in both the nucleus and cytoplasm [39,152,156]. In response to viral infection, DDX17 translocates from the nucleus to the cytoplasm where the vRNAs are located [37].

Recent work has implicated the exosome and components of a canonical cofactor complex in the direct recognition and degradation of specific vRNAs. RNAi screening revealed an antiviral role for exosome core components as well as the TRAMP components Mtr4 and Zcchc7 against three RNA viruses from distinct families, VSV, SINV, and RVFV, in both *Drosophila* and human cells [157]. Though TRAMP components are normally nucleolar, infection with these cytoplasmic viruses leads to the export of hMTR4 and hZCCHC7 to the cytoplasm, where they complex with the exosome and specifically bind viral mRNAs. Further study found that RVFV mRNA is destabilized by the exosome and hZCCHC7, and that the 3′ UTR of RVFV mRNA is sufficient to render a reporter RNA susceptible to exosomal degradation during viral infection. Cell biological studies showed that hZCCHC7 localizes to cytoplasmic punctae during viral infection, suggesting that it may be recruited to RNP granules for its antiviral function.

Taken together, these studies suggest that the exosome is a broad antiviral effector downstream of diverse sensors which bind distinct vRNAs to recruit the exosome for degradation (Figure 1). Though the exosome has been shown to degrade vRNA sensed by some of the proposed exosomal cofactors, such as Mtr4. Zcchc7, and ZAP, much work remains to describe the mechanism and exosomal involvement in antiviral restriction downstream of the other proteins.

9. Concluding Remarks

Increasing evidence suggests that the RNA decay machinery plays important roles in antiviral defense. This can involve either direct effects on vRNA stability or indirect regulation of the intracellular milieu. Furthermore, an emerging theme suggests that many RNA binding proteins can be repurposed from their endogenous roles in the nucleus to antiviral roles in the cytoplasm. Future studies are necessary to further elucidate how these RNA binding proteins recognize foreign RNAs and how they interface with the RNA decay machinery to restrict vRNA replication.

Acknowledgments: We thank E. Klinman, B. Hackett, and C. Sansone for critical reading of the manuscript. This work was supported by National Institute of Health grants R01AI074951, U54AI057168, and R01AI095500 to SC, and T32AI007324 to JMM. SC is a recipient of the Burroughs Wellcome Investigators in the Pathogenesis of Infectious Disease Award.

Conflicts of Interest: The authors declare no conflict of interest.

References

1.   Moon, S.L.; Wilusz, J. Cytoplasmic viruses: Rage against the (cellular RNA decay) machine. *PLoS Pathog.* **2013**, *9*, e1003762. [CrossRef] [PubMed]
2.   Decroly, E.; Ferron, F.; Lescar, J.; Canard, B. Conventional and unconventional mechanisms for capping viral mRNA. *Nat. Rev. Microbiol.* **2012**, *10*, 51–65. [CrossRef] [PubMed]

3. Moon, S.L.; Barnhart, M.D.; Wilusz, J. Inhibition and avoidance of mRNA degradation by RNA viruses. *Curr. Opin. Microbiol.* **2012**, *15*, 500–505. [CrossRef] [PubMed]

4. Ford, L.P.; Wilusz, J. 3′-Terminal RNA structures and poly(U) tracts inhibit initiation by a 3′-5′ exonuclease in vitro. *Nucleic Acids Res.* **1999**, *27*, 1159–1167. [CrossRef] [PubMed]

5. Iwakawa, H.-O.; Tajima, Y.; Taniguchi, T.; Kaido, M.; Mise, K.; Tomari, Y.; Taniguchi, H.; Okuno, T. Poly(A)-binding protein facilitates translation of an uncapped/nonpolyadenylated viral RNA by binding to the 3′ untranslated region. *J. Virol.* **2012**, *86*, 7836–7849. [CrossRef] [PubMed]

6. Barbalat, R.; Ewald, S.E.; Mouchess, M.L.; Barton, G.M. Nucleic acid recognition by the innate immune system. *Annu. Rev. Immunol.* **2011**, *29*, 185–214. [CrossRef] [PubMed]

7. Kato, H.; Takeuchi, O.; Sato, S.; Yoneyama, M.; Yamamoto, M.; Matsui, K.; Uematsu, S.; Jung, A.; Kawai, T.; Ishii, K.J.; et al. Differential roles of MDA5 and RIG-I helicases in the recognition of RNA viruses. *Nature* **2006**, *441*, 101–105. [CrossRef] [PubMed]

8. Hornung, V.; Ellegast, J.; Kim, S.; Brzózka, K.; Jung, A.; Kato, H.; Poeck, H.; Akira, S.; Conzelmann, K.-K.; Schlee, M.; Endres, S.; Hartmann, G. 5′-Triphosphate RNA is the ligand for RIG-I. *Science* **2006**, *314*, 994–997. [CrossRef] [PubMed]

9. Pichlmair, A.; Schulz, O.; Tan, C.-P.; Rehwinkel, J.; Kato, H.; Takeuchi, O.; Akira, S.; Way, M.; Schiavo, G.; Reis e Sousa, C. Activation of MDA5 requires higher-order RNA structures generated during virus infection. *J. Virol.* **2009**, *83*, 10761–10769. [CrossRef] [PubMed]

10. Gitlin, L.; Barchet, W.; Gilfillan, S.; Cella, M.; Beutler, B.; Flavell, R.A.; Diamond, M.S.; Colonna, M. Essential role of mda-5 in type I IFN responses to polyriboinosinic:polyribocytidylic acid and encephalomyocarditis picornavirus. *Proc. Natl. Acad. Sci. USA* **2006**, *103*, 8459–8464. [CrossRef] [PubMed]

11. Kato, H.; Takeuchi, O.; Mikamo-Satoh, E.; Hirai, R.; Kawai, T.; Matsushita, K.; Hiiragi, A.; Dermody, T.S.; Fujita, T.; Akira, S. Length-dependent recognition of double-stranded ribonucleic acids by retinoic acid-inducible gene-I and melanoma differentiation-associated gene 5. *J. Exp. Med.* **2008**, *205*, 1601–1610. [CrossRef] [PubMed]

12. Kawai, T.; Takahashi, K.; Sato, S.; Coban, C.; Kumar, H.; Kato, H.; Ishii, K.J.; Takeuchi, O.; Akira, S. IPS-1, an adaptor triggering RIG-I- and Mda5-mediated type I interferon induction. *Nat. Immunol.* **2005**, *6*, 981–988. [CrossRef] [PubMed]

13. Meylan, E.; Curran, J.; Hofmann, K.; Moradpour, D.; Binder, M.; Bartenschlager, R.; Tschopp, J. Cardif is an adaptor protein in the RIG-I antiviral pathway and is targeted by hepatitis C virus. *Nature* **2005**, *437*, 1167–1172. [CrossRef] [PubMed]

14. Seth, R.B.; Sun, L.; Ea, C.-K.; Chen, Z.J. Identification and characterization of MAVS, a mitochondrial antiviral signaling protein that activates NF-kappaB and IRF 3. *Cell* **2005**, *122*, 669–682. [CrossRef] [PubMed]

15. Xu, L.-G.; Wang, Y.-Y.; Han, K.-J.; Li, L.-Y.; Zhai, Z.; Shu, H.-B. VISA is an adapter protein required for virus-triggered IFN-beta signaling. *Mol. Cell* **2005**, *19*, 727–740. [CrossRef] [PubMed]

16. Bruns, A.M.; Leser, G.P.; Lamb, R.A.; Horvath, C.M. The innate immune sensor LGP2 activates antiviral signaling by regulating MDA5-RNA interaction and filament assembly. *Mol. Cell* **2014**, *55*, 771–781. [CrossRef] [PubMed]

17. Bruns, A.M.; Pollpeter, D.; Hadizadeh, N.; Myong, S.; Marko, J.F.; Horvath, C.M. ATP hydrolysis enhances RNA recognition and antiviral signal transduction by the innate immune sensor, laboratory of genetics and physiology 2 (LGP2). *J. Biol. Chem.* **2013**, *288*, 938–946. [CrossRef] [PubMed]

18. Komuro, A.; Horvath, C.M. RNA- and virus-independent inhibition of antiviral signaling by RNA helicase LGP2. *J. Virol.* **2006**, *80*, 12332–12342. [CrossRef] [PubMed]

19. Rothenfusser, S.; Goutagny, N.; DiPerna, G.; Gong, M.; Monks, B.G.; Schoenemeyer, A.; Yamamoto, M.; Akira, S.; Fitzgerald, K.A. The RNA helicase Lgp2 inhibits TLR-independent sensing of viral replication by Retinoic Acid-Inducible Gene-I. *J. Immunol.* **2005**, *175*, 5260–5268. [CrossRef] [PubMed]

20. Satoh, T.; Kato, H.; Kumagai, Y.; Yoneyama, M.; Sato, S.; Matsushita, K.; Tsujimura, T.; Fujita, T.; Akira, S.; Takeuchi, O. LGP2 is a positive regulator of RIG-I- and MDA5-mediated antiviral responses. *Proc. Natl. Acad. Sci. USA* **2010**, *107*, 1512–1517. [CrossRef] [PubMed]

21. Venkataraman, T.; Valdes, M.; Elsby, R.; Kakuta, S.; Caceres, G.; Saijo, S.; Iwakura, Y.; Barber, G.N. Loss of DExD/H Box RNA helicase LGP2 manifests disparate antiviral responses. *J. Immunol.* **2007**, *178*, 6444–6455. [CrossRef] [PubMed]

22. Yoneyama, M.; Kikuchi, M.; Matsumoto, K.; Imaizumi, T.; Miyagishi, M.; Taira, K.; Foy, E.; Loo, Y.-M.; Gale, M.; Akira, S.; et al. Shared and Unique Functions of the DExD/H-Box Helicases RIG-I, MDA5, and LGP2 in Antiviral Innate Immunity. *J. Immunol.* **2005**, *175*, 2851–2858. [CrossRef] [PubMed]

23. Billecocq, A.; Spiegel, M.; Vialat, P.; Kohl, A.; Weber, F.; Bouloy, M.; Haller, O. NSs protein of Rift Valley fever virus blocks interferon production by inhibiting host gene transcription. *J. Virol.* **2004**, *78*, 9798–9806. [CrossRef] [PubMed]

24. Bouloy, M.; Janzen, C.; Vialat, P.; Khun, H.; Pavlovic, J.; Huerre, M.; Haller, O. Genetic evidence for an interferon-antagonistic function of rift valley fever virus nonstructural protein NSs. *J. Virol.* **2001**, *75*, 1371–1377. [CrossRef] [PubMed]

25. Frolova, E.I.; Fayzulin, R.Z.; Cook, S.H.; Griffin, D.E.; Rice, C.M.; Frolov, I. Roles of nonstructural protein nsP2 and α/β interferons in determining the outcome of sindbis virus infection. *J. Virol.* **2002**, *76*, 11254–11264. [CrossRef] [PubMed]

26. Garmashova, N.; Gorchakov, R.; Frolova, E.; Frolov, I. Sindbis virus nonstructural protein nsP2 is cytotoxic and inhibits cellular transcription. *J. Virol.* **2006**, *80*, 5686–5696. [CrossRef] [PubMed]

27. Gorchakov, R.; Frolova, E.; Williams, B.R.G.; Rice, C.M.; Frolov, I. PKR-dependent and -independent mechanisms are involved in translational shutoff during Sindbis virus infection. *J. Virol.* **2004**, *78*, 8455–8467. [CrossRef] [PubMed]

28. Takeuchi, O.; Akira, S. RIG-I-like antiviral protein in flies. *Nat. Immunol.* **2008**, *9*, 1327–1328. [CrossRef] [PubMed]

29. Galiana-Arnoux, D.; Dostert, C.; Schneemann, A.; Hoffmann, J.A.; Imler, J.-L. Essential function in vivo for Dicer-2 in host defense against RNA viruses in drosophila. *Nat. Immunol.* **2006**, *7*, 590–597. [CrossRef] [PubMed]

30. Lee, Y.S.; Nakahara, K.; Pham, J.W.; Kim, K.; He, Z.; Sontheimer, E.J.; Carthew, R.W.; Abrahante, J.E.; Daul, A.L.; Li, M.; et al. Distinct Roles for Drosophila Dicer-1 and Dicer-2 in the siRNA/miRNA Silencing Pathways. *Cell* **2004**, *117*, 69–81. [CrossRef]

31. Wang, X.-H.; Aliyari, R.; Li, W.-X.; Li, H.-W.; Kim, K.; Carthew, R.; Atkinson, P.; Ding, S.-W. RNA interference directs innate immunity against viruses in adult Drosophila. *Science* **2006**, *312*, 452–454. [CrossRef] [PubMed]

32. Van Rij, R.P.; Saleh, M.-C.; Berry, B.; Foo, C.; Houk, A.; Antoniewski, C.; Andino, R. The RNA silencing endonuclease Argonaute 2 mediates specific antiviral immunity in Drosophila melanogaster. *Genes Dev.* **2006**, *20*, 2985–2995. [CrossRef] [PubMed]

33. Deddouche, S.; Matt, N.; Budd, A.; Mueller, S.; Kemp, C.; Galiana-Arnoux, D.; Dostert, C.; Antoniewski, C.; Hoffmann, J.A.; Imler, J.-L. The DExD/H-box helicase Dicer-2 mediates the induction of antiviral activity in drosophila. *Nat. Immunol.* **2008**, *9*, 1425–1432. [CrossRef] [PubMed]

34. Gregory, R.I.; Yan, K.-P.; Amuthan, G.; Chendrimada, T.; Doratotaj, B.; Cooch, N.; Shiekhattar, R. The Microprocessor complex mediates the genesis of microRNAs. *Nature* **2004**, *432*, 235–240. [CrossRef] [PubMed]

35. Mori, M.; Triboulet, R.; Mohseni, M.; Schlegelmilch, K.; Shrestha, K.; Camargo, F.D.; Gregory, R.I. Hippo signaling regulates microprocessor and links cell-density-dependent miRNA biogenesis to cancer. *Cell* **2014**, *156*, 893–906. [CrossRef] [PubMed]

36. Suzuki, H.I.; Yamagata, K.; Sugimoto, K.; Iwamoto, T.; Kato, S.; Miyazono, K. Modulation of microRNA processing by p53. *Nature* **2009**, *460*, 529–533. [CrossRef] [PubMed]

37. Moy, R.H.; Cole, B.S.; Yasunaga, A.; Gold, B.; Shankarling, G.; Varble, A.; Molleston, J.M.; tenOever, B.R.; Lynch, K.W.; Cherry, S. Stem-loop recognition by DDX17 facilitates miRNA processing and antiviral defense. *Cell* **2014**, *158*, 764–777. [CrossRef] [PubMed]

38. Zhang, Z.; Kim, T.; Bao, M.; Facchinetti, V.; Jung, S.Y.; Ghaffari, A.A.; Qin, J.; Cheng, G.; Liu, Y.-J. DDX1, DDX21, and DHX36 helicases form a complex with the adaptor molecule TRIF to sense dsRNA in dendritic cells. *Immunity* **2011**, *34*, 866–878. [CrossRef] [PubMed]

39. Miyashita, M.; Oshiumi, H.; Matsumoto, M.; Seya, T. DDX60, a DEXD/H box helicase, is a novel antiviral factor promoting RIG-I-Like Receptor-mediated signaling. *Mol. Cell. Biol.* **2011**, *31*, 3802–3819. [CrossRef] [PubMed]

40. Laredj, L.N.; Beard, P. Adeno-associated virus activates an innate immune response in normal human cells but not in osteosarcoma cells. *J. Virol.* **2011**, *85*, 13133–13143. [CrossRef] [PubMed]

41. Kawai, T.; Akira, S. The roles of TLRs, RLRs and NLRs in pathogen recognition. *Int. Immunol.* **2009**, *21*, 317–337. [CrossRef] [PubMed]

42. Weber, C.; Müller, C.; Podszuweit, A.; Montino, C.; Vollmer, J.; Forsbach, A. Toll-like receptor (TLR) 3 immune modulation by unformulated small interfering RNA or DNA and the role of CD14 (in TLR-mediated effects). *Immunology* **2012**, *136*, 64–77. [CrossRef] [PubMed]

43. Clemens, M.J.; Hershey, J.W.; Hovanessian, A.C.; Jacobs, B.C.; Katze, M.G.; Kaufman, R.J.; Lengyel, P.; Samuel, C.E.; Sen, G.C.; Williams, B.R. PKR: Proposed nomenclature for the RNA-dependent protein kinase induced by interferon. *J. Interf. Res.* **1993**, *13*, 241. [CrossRef] [PubMed]

44. Gale, M.; Katze, M.G. Molecular mechanisms of interferon resistance mediated by viral-directed inhibition of PKR, the interferon-induced protein kinase. *Pharmacol. Ther.* **1998**, *78*, 29–46. [CrossRef]

45. Hinnebusch, A.G. The eIF-2α kinases: Regulators of protein synthesis in starvation and stress. *Semin. Cell. Biol.* **1994**, *5*, 417–426. [CrossRef] [PubMed]

46. Meurs, E.; Chong, K.; Galabru, J.; Thomas, N.S.B.; Kerr, I.M.; Williams, B.R.G.; Hovanessian, A.G. Molecular cloning and characterization of the human double-stranded RNA-activated protein kinase induced by interferon. *Cell* **1990**, *62*, 379–390. [CrossRef]

47. Chacko, M.S.; Adamo, M.L. Double-stranded RNA decreases IGF-I gene expression in a protein kinase R-Dependent, but type I interferon-independent, mechanism in C6 Rat Glioma Cells. *Endocrinol. Endocr. Soc.* **2011**. [CrossRef] [PubMed]

48. Balachandran, S.; Roberts, P.C.; Brown, L.E.; Truong, H.; Pattnaik, A.K.; Archer, D.R.; Barber, G.N. Essential role for the dsRNA-dependent protein kinase PKR in innate immunity to viral infection. *Immunity* **2000**, *13*, 129–141. [CrossRef]

49. Short, J.A.L. Viral evasion of interferon stimulated genes. *Biosci. Horiz.* **2009**, *2*, 212–224. [CrossRef]

50. Black, T.L.; Safer, B.; Hovanessian, A.; Katze, M.G. The cellular 68,000-Mr protein kinase is highly autophosphorylated and activated yet significantly degraded during poliovirus infection: Implications for translational regulation. *J. Virol.* **1989**, *63*, 2244–2251. [PubMed]

51. Ikegami, T.; Narayanan, K.; Won, S.; Kamitani, W.; Peters, C.J.; Makino, S. Rift Valley fever virus NSs protein promotes post-transcriptional downregulation of protein kinase PKR and inhibits eIF2alpha phosphorylation. *PLoS Pathog.* **2009**, *5*, e1000287. [CrossRef] [PubMed]

52. Kainulainen, M.; Lau, S.; Samuel, C.E.; Hornung, V.; Weber, F. NSs virulence factor of rift valley fever virus engages the F-Box proteins FBXW11 and β-TRCP1 to degrade the antiviral protein Kinase PKR. *J. Virol.* **2016**, *90*, 6140–6147. [CrossRef] [PubMed]

53. Garaigorta, U.; Chisari, F.V. Hepatitis C virus blocks interferon effector function by inducing protein kinase R phosphorylation. *Cell. Host Microbe* **2009**, *6*, 513–522. [CrossRef] [PubMed]

54. Kim, J.H.; Park, S.M.; Park, J.H.; Keum, S.J.; Jang, S.K. eIF2A mediates translation of hepatitis C viral mRNA under stress conditions. *EMBO J.* **2011**, *30*, 2454–2464. [CrossRef] [PubMed]

55. Robert, F.; Kapp, L.D.; Khan, S.N.; Acker, M.G.; Kolitz, S.; Kazemi, S.; Kaufman, R.J.; Merrick, W.C.; Koromilas, A.E.; Lorsch, J.R.; et al. Initiation of protein synthesis by hepatitis C virus is refractory to reduced eIF2.GTP.Met-tRNA(i)(Met) ternary complex availability. *Mol. Biol. Cell* **2006**, *17*, 4632–4644. [CrossRef] [PubMed]

56. Schneider, R.J.; Mohr, I. Translation initiation and viral tricks. *Trends Biochem.Sci.* **2003**, *28*, 130–136. [CrossRef]

57. Sen, A.; Pruijssers, A.J.; Dermody, T.S.; García-Sastre, A.; Greenberg, H.B. The early interferon response to rotavirus is regulated by PKR and depends on MAVS/IPS-1, RIG-I, MDA-5, and IRF3. *J. Virol.* **2011**, *85*, 3717–3732. [CrossRef] [PubMed]

58. Zhang, P.; Langland, J.O.; Jacobs, B.L.; Samuel, C.E. Protein kinase PKR-dependent activation of mitogen-activated protein kinases occurs through mitochondrial adapter IPS-1 and is antagonized by vaccinia virus E3L. *J. Virol.* **2009**, *83*, 5718–5725. [CrossRef] [PubMed]

59. Hovanessian, A.G.; Laurent, A.G.; Chebath, J.; Galabru, J.; Robert, N.; Svab, J. Identification of 69-kd and 100-kd forms of 2–5A synthetase in interferon-treated human cells by specific monoclonal antibodies. *EMBO J.* **1987**, *6*, 1273–1280. [PubMed]

60. Jensen, S.; Thomsen, A.R. Sensing of RNA viruses: A review of innate immune receptors involved in recognizing RNA virus invasion. *J. Virol.* **2012**, *86*, 2900–2910. [CrossRef] [PubMed]

61. Floyd-Smith, G.; Slattery, E.; Lengyel, P. Interferon action: RNA cleavage pattern of a (2′-5′)oligoadenylate–dependent endonuclease. *Science* **1981**, *212*, 1030–1032. [CrossRef] [PubMed]

62. Wreschner, D.H.; McCauley, J.W.; Skehel, J.J.; Kerr, I.M. Interferon action—Sequence specificity of the ppp(A2′p)nA-dependent ribonuclease. *Nature* **1981**, *289*, 414–417. [CrossRef] [PubMed]

63. Silverman, R.H. Viral encounters with 2′,5′-oligoadenylate synthetase and RNase L during the interferon antiviral response. *J. Virol.* **2007**, *81*, 12720–12729. [CrossRef] [PubMed]

64. Malathi, K.; Dong, B.; Gale, M.; Silverman, R.H. Small self-RNA generated by RNase L amplifies antiviral innate immunity. *Nature* **2007**, *448*, 816–819. [CrossRef] [PubMed]

65. Castelli, J.C.; Hassel, B.A.; Wood, K.A.; Li, X.L.; Amemiya, K.; Dalakas, M.C.; Torrence, P.F.; Youle, R.J. A study of the interferon antiviral mechanism: Apoptosis activation by the 2–5A system. *J. Exp. Med.* **1997**, *186*, 967–972. [CrossRef] [PubMed]

66. Brennan-Laun, S.E.; Ezelle, H.J.; Li, X.-L.; Hassel, B.A. RNase-L control of cellular mRNAs: Roles in biologic functions and mechanisms of substrate targeting. *J. Interf. Cytokine Res.* **2014**, *34*, 275–288. [CrossRef] [PubMed]

67. Shapiro, J.S.; Schmid, S.; Aguado, L.C.; Sabin, L.R.; Yasunaga, A.; Shim, J.V.; Sachs, D.; Cherry, S.; tenOever, B.R. Drosha as an interferon-independent antiviral factor. *Proc. Natl. Acad. Sci. USA* **2014**, *111*, 7108–7113. [CrossRef] [PubMed]

68. Laguette, N.; Sobhian, B.; Casartelli, N.; Ringeard, M.; Chable-Bessia, C.; Ségéral, E.; Yatim, A.; Emiliani, S.; Schwartz, O.; Benkirane, M. SAMHD1 is the dendritic- and myeloid-cell-specific HIV-1 restriction factor counteracted by Vpx. *Nature* **2011**, *474*, 654–657. [CrossRef] [PubMed]

69. Goldstone, D.C.; Ennis-Adeniran, V.; Hedden, J.J.; Groom, H.C.T.; Rice, G.I.; Christodoulou, E.; Walker, P.A.; Kelly, G.; Haire, L.F.; Yap, M.W.; et al. HIV-1 restriction factor SAMHD1 is a deoxynucleoside triphosphate triphosphohydrolase. *Nature* **2011**, *480*, 379–382. [CrossRef] [PubMed]

70. Powell, R.D.; Holland, P.J.; Hollis, T.; Perrino, F.W. Aicardi-Goutieres syndrome gene and HIV-1 restriction factor SAMHD1 is a dGTP-regulated deoxynucleotide triphosphohydrolase. *J. Biol. Chem.* **2011**, *286*, 43596–43600. [CrossRef] [PubMed]

71. Beloglazova, N.; Flick, R.; Tchigvintsev, A.; Brown, G.; Popovic, A.; Nocek, B.; Yakunin, A.F. Nuclease activity of the human SAMHD1 protein implicated in the Aicardi-Goutieres syndrome and HIV-1 restriction. *J. Biol. Chem.* **2013**, *288*, 8101–8110. [CrossRef] [PubMed]

72. Allmang, C.; Kufel, J.; Chanfreau, G.; Mitchell, P.; Petfalski, E.; Tollervey, D. Functions of the exosome in rRNA, snoRNA and snRNA synthesis. *EMBO J.* **1999**, *18*, 5399–5410. [CrossRef] [PubMed]

73. Milligan, L.; Torchet, C.; Allmang, C.; Shipman, T.; Tollervey, D. A nuclear surveillance pathway for mRNAs with defective polyadenylation. *Mol. Cell. Biol.* **2005**, *25*, 9996–10004. [CrossRef] [PubMed]

74. Kadaba, S. Nuclear surveillance and degradation of hypomodified initiator tRNAMet in S. cerevisiae. *Genes Dev.* **2004**, *18*, 1227–1240. [CrossRef] [PubMed]

75. Isken, O.; Maquat, L.E. Quality control of eukaryotic mRNA: Safeguarding cells from abnormal mRNA function. *Genes Dev.* **2007**, *21*, 1833–1856. [CrossRef] [PubMed]

76. Keene, J.D. RNA regulons: Coordination of post-transcriptional events. *Nat. Rev. Genet.* **2007**, *8*, 533–543. [CrossRef] [PubMed]

77. Hopkins, K.C.; Tartell, M.A.; Herrmann, C.; Hackett, B.A.; Taschuk, F.; Panda, D.; Menghani, S.V.; Sabin, L.R.; Cherry, S. Virus-induced translational arrest through 4EBP1/2-dependent decay of 5′-TOP mRNAs restricts viral infection. *Proc. Natl. Acad. Sci. USA* **2015**, *112*, 201418805. [CrossRef] [PubMed]

78. Mukherjee, D.; Gao, M.; O'Connor, J.P.; Raijmakers, R.; Pruijn, G.; Lutz, C.S.; Wilusz, J. The mammalian exosome mediates the efficient degradation of mRNAs that contain AU-rich elements. *EMBO J.* **2002**, *21*, 165–174. [CrossRef] [PubMed]

79. Chen, C.Y.; Gherzi, R.; Ong, S.E.; Chan, E.L.; Raijmakers, R.; Pruijn, G.J.; Stoecklin, G.; Moroni, C.; Mann, M.; Karin, M. AU binding proteins recruit the exosome to degrade ARE-containing mRNAs. *Cell* **2001**, *107*, 451–464. [CrossRef]

80. Lejeune, F.; Li, X.; Maquat, L.E. Nonsense-mediated mRNA decay in mammalian cells involves decapping, deadenylating, and exonucleolytic activities. *Mol. Cell* **2003**, *12*, 675–687. [CrossRef]

81. Baker, K.E.; Parker, R. Nonsense-mediated mRNA decay: Terminating erroneous gene expression. *Curr. Opin. Cell Biol.* **2004**, *16*, 293–299. [CrossRef] [PubMed]

82. Balistreri, G.; Horvath, P.; Schweingruber, C.; Zünd, D.; McInerney, G.; Merits, A.; Mühlemann, O.; Azzalin, C.; Helenius, A. The host nonsense-mediated mRNA decay pathway restricts Mammalian RNA virus replication. *Cell. Host Microbe* **2014**, *16*, 403–411. [CrossRef] [PubMed]

83. Garcia, D.; Garcia, S.; Voinnet, O. Nonsense-mediated decay serves as a general viral restriction mechanism in plants. *Cell. Host Microbe* **2014**, *16*, 391–402. [CrossRef] [PubMed]
84. Nakano, K.; Ando, T.; Yamagishi, M.; Yokoyama, K.; Ishida, T.; Ohsugi, T.; Tanaka, Y.; Brighty, D.W.; Watanabe, T. Viral interference with host mRNA surveillance, the nonsense-mediated mRNA decay (NMD) pathway, through a new function of HTLV-1 Rex: Implications for retroviral replication. *Microbes Infect.* **2013**, *15*, 491–505. [CrossRef] [PubMed]
85. Ramage, H.R.; Kumar, G.R.; Verschueren, E.; Johnson, J.R.; Von Dollen, J.; Johnson, T.; Newton, B.; Shah, P.; Horner, J.; Krogan, N.J.; et al. A combined proteomics/genomics approach links hepatitis C virus infection with nonsense-mediated mRNA decay. *Mol. Cell* **2015**, *57*, 329–340. [CrossRef] [PubMed]
86. Parker, R.; Sheth, U. P Bodies and the Control of mRNA Translation and Degradation. *Mol. Cell* **2007**, *25*, 635–646. [CrossRef] [PubMed]
87. Huarte, J.; Stutz, A.; O'Connell, M.L.; Gubler, P.; Belin, D.; Darrow, A.L.; Strickland, S.; Vassalli, J.-D. Transient translational silencing by reversible mRNA deadenylation. *Cell* **1992**, *69*, 1021–1030. [CrossRef]
88. Muhlrad, D.; Decker, C.J.; Parker, R. Deadenylation of the unstable mRNA encoded by the yeast MFA2 gene leads to decapping followed by $5'$–>$3'$ digestion of the transcript. *Genes Dev.* **1994**, *8*, 855–866. [CrossRef] [PubMed]
89. Hsu, C.L.; Stevens, A. Yeast cells lacking $5'$–>$3'$ exoribonuclease 1 contain mRNA species that are poly(A) deficient and partially lack the $5'$ cap structure. *Mol. Cell. Biol.* **1993**, *13*, 4826–4835. [CrossRef] [PubMed]
90. Brannan, K.; Kim, H.; Erickson, B.; Glover-Cutter, K.; Kim, S.; Fong, N.; Kiemele, L.; Hansen, K.; Davis, R.; Lykke-Andersen, J.; et al. mRNA decapping factors and the exonuclease Xrn2 function in widespread premature termination of RNA polymerase II transcription. *Mol. Cell* **2012**, *46*, 311–324. [CrossRef] [PubMed]
91. Li, Y.; Song, M.; Kiledjian, M. Differential utilization of decapping enzymes in mammalian mRNA decay pathways. *RNA* **2011**, *17*, 419–428. [CrossRef] [PubMed]
92. Sheth, U.; Parker, R. Decapping and decay of messenger RNA occur in cytoplasmic processing bodies. *Science* **2003**, *300*, 805–808. [CrossRef] [PubMed]
93. Brengues, M.; Teixeira, D.; Parker, R. Movement of Eukaryotic mRNAs between polysomes and cytoplasmic processing bodies. *Science* **2005**, *310*, 486–489. [CrossRef] [PubMed]
94. Ayache, J.; Bénard, M.; Ernoult-Lange, M.; Minshall, N.; Standart, N.; Kress, M.; Weil, D. P-body assembly requires DDX6 repression complexes rather than decay or Ataxin2/2L complexes. *Mol. Biol. Cell* **2015**, *26*, 2579–2595. [CrossRef] [PubMed]
95. Kedersha, N.; Stoecklin, G.; Ayodele, M.; Yacono, P.; Lykke-Andersen, J.; Fritzler, M.J.; Scheuner, D.; Kaufman, R.J.; Golan, D.E.; Anderson, P. Stress granules and processing bodies are dynamically linked sites of mRNP remodeling. *J. Cell Biol.* **2005**, *169*, 871–884. [CrossRef] [PubMed]
96. Wilczynska, A.; Aigueperse, C.; Kress, M.; Dautry, F.; Weil, D. The translational regulator CPEB1 provides a link between dcp1 bodies and stress granules. *J. Cell. Sci.* **2005**, *118*, 981–992. [CrossRef] [PubMed]
97. Eulalio, A.; Behm-Ansmant, I.; Schweizer, D.; Izaurralde, E. P-body formation is a consequence, not the cause, of RNA-mediated gene silencing. *Mol. Cell. Biol.* **2007**, *27*, 3970–3981. [CrossRef] [PubMed]
98. Teixeira, D.; Sheth, U.; Valencia-Sanchez, M.A.; Brengues, M.; Parker, R. Processing bodies require RNA for assembly and contain nontranslating mRNAs. *RNA* **2005**, *11*, 371–382. [CrossRef] [PubMed]
99. Jones, D.M.; Domingues, P.; Targett-Adams, P.; McLauchlan, J. Comparison of U2OS and Huh-7 cells for identifying host factors that affect hepatitis C virus RNA replication. *J. Gen. Virol.* **2010**, *91*, 2238–2248. [CrossRef] [PubMed]
100. Moon, S.L.; Anderson, J.R.; Kumagai, Y.; Wilusz, C.J.; Akira, S.; Khromykh, A.A.; Wilusz, J. A noncoding RNA produced by arthropod-borne flaviviruses inhibits the cellular exoribonuclease XRN1 and alters host mRNA stability. *RNA* **2012**, *18*, 2029–2040. [CrossRef] [PubMed]
101. Silva, P.A.G.C.; Pereira, C.F.; Dalebout, T.J.; Spaan, W.J.M.; Bredenbeek, P.J. An RNA pseudoknot is required for production of yellow fever virus subgenomic RNA by the host nuclease XRN1. *J. Virol.* **2010**, *84*, 11395–11406. [CrossRef] [PubMed]
102. Dougherty, J.D.; White, J.P.; Lloyd, R.E. Poliovirus-mediated disruption of cytoplasmic processing bodies. *J. Virol.* **2011**, *85*, 64–75. [CrossRef] [PubMed]
103. Hopkins, K.C.; McLane, L.M.; Maqbool, T.; Panda, D.; Gordesky-Gold, B.; Cherry, S. A genome-wide RNAi screen reveals that mRNA decapping restricts bunyaviral replication by limiting the pools of Dcp2-accessible targets for cap-snatching. *Genes Dev.* **2013**, *27*, 1511–1525. [CrossRef] [PubMed]

104. Emara, M.M.; Brinton, M.A. Interaction of TIA-1/TIAR with West Nile and dengue virus products in infected cells interferes with stress granule formation and processing body assembly. *Proc. Natl. Acad. Sci. USA* **2007**, *104*, 9041–9046. [CrossRef] [PubMed]

105. Chahar, H.S.; Chen, S.; Manjunath, N. P-body components LSM1, GW182, DDX3, DDX6 and XRN1 are recruited to WNV replication sites and positively regulate viral replication. *Virology* **2013**, *436*, 1–7. [CrossRef] [PubMed]

106. Ward, A.M.; Bidet, K.; Yinglin, A.; Ler, S.G.; Hogue, K.; Blackstock, W.; Gunaratne, J.; Garcia-Blanco, M.A. Quantitative mass spectrometry of DENV-2 RNA-interacting proteins reveals that the DEAD-box RNA helicase DDX6 binds the DB1 and DB2 3′ UTR structures. *RNA Biol.* **2011**, *8*, 1173–1186. [CrossRef] [PubMed]

107. Schneider, C.; Tollervey, D. Threading the barrel of the RNA exosome. *Trends Biochem. Sci.* **2013**, *38*, 485–493. [CrossRef] [PubMed]

108. Liu, Q.; Greimann, J.C.; Lima, C.D. Reconstitution, activities, and structure of the eukaryotic RNA exosome. *Cell* **2006**, *127*, 1223–1237. [CrossRef] [PubMed]

109. Raijmakers, R.; Egberts, W.V.; van Venrooij, W.J.; Pruijn, G.J.M. Protein-Protein Interactions between Human Exosome Components Support the Assembly of RNase PH-type Subunits into a Six-membered PNPase-like Ring. *J. Mol. Biol.* **2002**, *323*, 653–663. [CrossRef]

110. Makino, D.L.; Baumgärtner, M.; Conti, E. Crystal structure of an RNA-bound 11-subunit eukaryotic exosome complex. *Nature* **2013**, *495*, 70–75. [CrossRef] [PubMed]

111. Lin, W.-J.; Duffy, A.; Chen, C.-Y. Localization of AU-rich element-containing mRNA in cytoplasmic granules containing exosome subunits. *J. Biol. Chem.* **2007**, *282*, 19958–19968. [CrossRef] [PubMed]

112. Graham, A.C.; Kiss, D.L.; Andrulis, E.D. Differential distribution of exosome subunits at the nuclear lamina and in cytoplasmic foci. *Mol. Biol. Cell* **2006**, *17*, 1399–1409. [CrossRef] [PubMed]

113. Zurla, C.; Lifland, A.W.; Santangelo, P.J. Characterizing mRNA interactions with RNA granules during translation initiation inhibition. *PLoS ONE* **2011**, *6*, e19727. [CrossRef] [PubMed]

114. Dziembowski, A.; Lorentzen, E.; Conti, E.; Séraphin, B. A single subunit, Dis3, is essentially responsible for yeast exosome core activity. *Nat. Struct. Mol. Biol.* **2007**, *14*, 15–22. [CrossRef] [PubMed]

115. Mitchell, P.; Petfalski, E.; Shevchenko, A.; Mann, M.; Tollervey, D. The exosome: A conserved eukaryotic RNA processing complex containing multiple 3′->5′ exoribonucleases. *Cell* **1997**, *91*, 457–466. [CrossRef]

116. Allmang, C.; Petfalski, E.; Podtelejnikov, A.; Mann, M.; Tollervey, D.; Mitchell, P. The yeast exosome and human PM-Scl are related complexes of 3′->5′ exonucleases. *Genes Dev.* **1999**, *13*, 2148–2158. [CrossRef] [PubMed]

117. Houseley, J.; Tollervey, D. The many pathways of RNA degradation. *Cell* **2009**, *136*, 763–776. [CrossRef] [PubMed]

118. Tomecki, R.; Kristiansen, M.S.; Lykke-Andersen, S.; Chlebowski, A.; Larsen, K.M.; Szczesny, R.J.; Drazkowska, K.; Pastula, A.; Andersen, J.S.; Stepien, P.P.; et al. The human core exosome interacts with differentially localized processive RNases: hDIS3 and hDIS3L. *EMBO J.* **2010**, *29*, 2342–2357. [CrossRef] [PubMed]

119. Chang, H.-M.; Triboulet, R.; Thornton, J.E.; Gregory, R.I. A role for the Perlman syndrome exonuclease Dis3l2 in the Lin28-let-7 pathway. *Nature* **2013**, *497*, 244–248. [CrossRef] [PubMed]

120. Thomas, M.P.; Liu, X.; Whangbo, J.; McCrossan, G.; Sanborn, K.B.; Basar, E.; Walch, M.; Lieberman, J. Apoptosis triggers specific, rapid, and global mRNA Decay with 3′ uridylated intermediates degraded by DIS3L2. *Cell. Rep.* **2015**, *11*, 1079–1089. [CrossRef] [PubMed]

121. Ustianenko, D.; Hrossova, D.; Potesil, D.; Chalupnikova, K.; Hrazdilova, K.; Pachernik, J.; Cetkovska, K.; Uldrijan, S.; Zdrahal, Z.; Vanacova, S. Mammalian DIS3L2 exoribonuclease targets the uridylated precursors of let-7 miRNAs. *RNA* **2013**, *19*, 1632–1638. [CrossRef] [PubMed]

122. Houseley, J.; LaCava, J.; Tollervey, D. RNA-quality control by the exosome. *Nat. Rev. Mol. Cell Biol.* **2006**, *7*, 529–539. [CrossRef] [PubMed]

123. De la Cruz, J.; Kressler, D.; Tollervey, D.; Linder, P. Dob1p (Mtr4p) is a putative ATP-dependent RNA helicase required for the 3′ end formation of 5.8S rRNA in Saccharomyces cerevisiae. *EMBO J.* **1998**, *17*, 1128–1140. [CrossRef] [PubMed]

124. Toh-E, A.; Guerry, P.; Wickner, R.B. Chromosomal superkiller mutants of Saccharomyces cerevisiae. *J. Bacteriol.* **1978**, *136*, 1002–1007. [PubMed]

125. Masison, D.C.; Blanc, A.; Ribas, J.C.; Carroll, K.; Sonenberg, N.; Wickner, R.B. Decoying the cap- mRNA degradation system by a double-stranded RNA virus and poly(A)- mRNA surveillance by a yeast antiviral system. *Mol. Cell. Biol.* **1995**, *15*, 2763–2771. [CrossRef] [PubMed]

126. Brown, J.T.; Bai, X.; Johnson, A.W. The yeast antiviral proteins Ski2p, Ski3p, and Ski8p exist as a complex in vivo. *RNA* **2000**, *6*, 449–457. [CrossRef] [PubMed]

127. Araki, Y.; Takahashi, S.; Kobayashi, T.; Kajiho, H.; Hoshino, S.; Katada, T. Ski7p G protein interacts with the exosome and the Ski complex for 3′-to-5′ mRNA decay in yeast. *EMBO J.* **2001**, *20*, 4684–4693. [CrossRef] [PubMed]

128. Jacobs Anderson, J.S.; Parker, R. The 3′ to 5′ degradation of yeast mRNAs is a general mechanism for mRNA turnover that requires the SKI2 DEVH box protein and 3′ to 5′ exonucleases of the exosome complex. *EMBO J.* **1998**, *17*, 1497–1506. [CrossRef] [PubMed]

129. Van Hoof, A.; Frischmeyer, P.A.; Dietz, H.C.; Parker, R. Exosome-mediated recognition and degradation of mRNAs lacking a termination codon. *Science* **2002**, *295*, 2262–2264. [CrossRef] [PubMed]

130. Horikawa, W.; Endo, K.; Wada, M.; Ito, K. Mutations in the G-domain of Ski7 cause specific dysfunction in non-stop decay. *Sci. Rep.* **2016**, *6*, 29295. [CrossRef] [PubMed]

131. Orban, T.I.; Izaurralde, E. Decay of mRNAs targeted by RISC requires XRN1, the Ski complex, and the exosome. *RNA* **2005**, *11*, 459–469. [CrossRef] [PubMed]

132. Zhu, B.; Mandal, S.S.; Pham, A.-D.; Zheng, Y.; Erdjument-Bromage, H.; Batra, S.K.; Tempst, P.; Reinberg, D. The human PAF complex coordinates transcription with events downstream of RNA synthesis. *Genes Dev.* **2005**, *19*, 1668–1673. [CrossRef] [PubMed]

133. Eckard, S.C.; Rice, G.I.; Fabre, A.; Badens, C.; Gray, E.E.; Hartley, J.L.; Crow, Y.J.; Stetson, D.B. The SKIV2L RNA exosome limits activation of the RIG-I-like receptors. *Nat. Immunol.* **2014**, *15*, 839–845. [CrossRef] [PubMed]

134. Cai, X.; Chiu, Y.-H.; Chen, Z. J. The cGAS-cGAMP-STING Pathway of Cytosolic DNA Sensing and Signaling. *Mol. Cell* **2014**, *54*, 289–296. [CrossRef] [PubMed]

135. Crow, Y.J.; Rehwinkel, J. Aicardi-Goutieres syndrome and related phenotypes: Linking nucleic acid metabolism with autoimmunity. *Hum. Mol. Genet.* **2009**, *18*, R130–R136. [CrossRef] [PubMed]

136. Stetson, D.B.; Ko, J.S.; Heidmann, T.; Medzhitov, R. Trex1 prevents cell-intrinsic initiation of autoimmunity. *Cell* **2008**, *134*, 587–598. [CrossRef] [PubMed]

137. LaCava, J.; Houseley, J.; Saveanu, C.; Petfalski, E.; Thompson, E.; Jacquier, A.; Tollervey, D. RNA Degradation by the Exosome Is Promoted by a Nuclear Polyadenylation Complex. *Cell* **2005**, *121*, 713–724. [CrossRef] [PubMed]

138. Wyers, F.; Rougemaille, M.; Badis, G.; Rousselle, J.-C.; Dufour, M.-E.; Boulay, J.; Régnault, B.; Devaux, F.; Namane, A.; Séraphin, B. Cryptic Pol II Transcripts Are Degraded by a Nuclear Quality Control Pathway Involving a New Poly(A) Polymerase. *Cell* **2005**, *121*, 725–737. [CrossRef] [PubMed]

139. Jackson, R.N.; Klauer, A.A.; Hintze, B.J.; Robinson, H.; van Hoof, A.; Johnson, S.J. The crystal structure of Mtr4 reveals a novel arch domain required for rRNA processing. *EMBO J.* **2010**, *29*, 2205–2216. [CrossRef] [PubMed]

140. Schmidt, K.; Xu, Z.; Mathews, D.H.; Butler, J.S. Air proteins control differential TRAMP substrate specificity for nuclear RNA surveillance. *RNA* **2012**, *18*, 1934–1945. [CrossRef] [PubMed]

141. Paolo, S.S.; Vanacova, S.; Schenk, L.; Scherrer, T.; Blank, D.; Keller, W.; Gerber, A.P. Distinct Roles of Non-Canonical Poly(A) Polymerases in RNA Metabolism. *PLoS Genet.* **2009**, *5*, e1000555. [CrossRef] [PubMed]

142. Li, Z.; Reimers, S.; Pandit, S.; Deutscher, M.P.; Bradford, M.; Carpousis, A.; Vanzo, N.; Raynal, L.; Coburn, G.; Mackie, G.; et al. RNA quality control: Degradation of defective transfer RNA. *EMBO J.* **2002**, *21*, 1132–1138. [CrossRef] [PubMed]

143. Deutscher, M.P. Degradation of RNA in bacteria: Comparison of mRNA and stable RNA. *Nucleic Acids Res.* **2006**, *34*, 659–666. [CrossRef] [PubMed]

144. Thoms, M.; Thomson, E.; Baßler, J.; Gnädig, M.; Griesel, S.; Hurt, E. The exosome is recruited to RNA substrates through specific adaptor proteins. *Cell* **2015**, *162*, 1029–1038. [CrossRef] [PubMed]

145. Lubas, M.; Christensen, M.S.; Kristiansen, M.S.; Domanski, M.; Falkenby, L.G.; Lykke-Andersen, S.; Andersen, J.S.; Dziembowski, A.; Jensen, T.H. Interaction profiling identifies the human nuclear exosome targeting complex. *Mol. Cell* **2011**, *43*, 624–637. [CrossRef] [PubMed]

146. Fasken, M.B.; Leung, S.W.; Banerjee, A.; Kodani, M.O.; Chavez, R.; Bowman, E.A.; Purohit, M.K.; Rubinson, M.E.; Rubinson, E.H.; Corbett, A.H. Air1 zinc knuckles 4 and 5 and a conserved IWRXY motif are critical for the function and integrity of the Trf4/5-Air1/2-Mtr4 polyadenylation (TRAMP) RNA quality control complex. *J. Biol. Chem.* **2011**, *286*, 37429–37445. [CrossRef] [PubMed]

147. Shcherbik, N.; Wang, M.; Lapik, Y.R.; Srivastava, L.; Pestov, D.G. Polyadenylation and degradation of incomplete RNA polymerase I transcripts in mammalian cells. *EMBO Rep.* **2010**, *11*, 106–111. [CrossRef] [PubMed]

148. Lubas, M.; Andersen, P.R.; Schein, A.; Dziembowski, A.; Kudla, G.; Jensen, T.H. The human nuclear exosome targeting complex is loaded onto newly synthesized RNA to direct early ribonucleolysis. *Cell Rep.* **2015**, *10*, 178–192. [CrossRef] [PubMed]

149. Andersen, P.R.; Domanski, M.; Kristiansen, M.S.; Storvall, H.; Ntini, E.; Verheggen, C.; Schein, A.; Bunkenborg, J.; Poser, I.; Hallais, M.; et al. The human cap-binding complex is functionally connected to the nuclear RNA exosome. *Nat. Struct. Mol. Biol.* **2013**, *20*, 1367–1376. [CrossRef] [PubMed]

150. Dorweiler, J.E.; Ni, T.; Zhu, J.; Munroe, S.H.; Anderson, J.T. Certain adenylated non-coding RNAs, including 5′ leader sequences of primary microrna transcripts, accumulate in mouse cells following depletion of the RNA Helicase MTR4. *PLoS ONE* **2014**, *9*, e99430. [CrossRef] [PubMed]

151. Chen, G.; Guo, X.; Lv, F.; Xu, Y.; Gao, G. p72 DEAD box RNA helicase is required for optimal function of the zinc-finger antiviral protein. *Proc. Natl. Acad. Sci. USA* **2008**, *105*, 4352–4357. [CrossRef] [PubMed]

152. Liang, G.; Liu, G.; Kitamura, K.; Wang, Z.; Chowdhury, S.; Monjurul, A.M.; Wakae, K.; Koura, M.; Shimadu, M.; Kinoshita, K.; et al. TGF-β suppression of HBV RNA through AID-dependent recruitment of an RNA exosome complex. *PLOS Pathog.* **2015**, *11*, e1004780. [CrossRef] [PubMed]

153. Guo, X.; Carroll, J.-W.N.; Macdonald, M.R.; Goff, S.P.; Gao, G. The zinc finger antiviral protein directly binds to specific viral mRNAs through the CCCH zinc finger motifs. *J. Virol.* **2004**, *78*, 12781–12787. [CrossRef] [PubMed]

154. Guo, X.; Ma, J.; Sun, J.; Gao, G. The zinc-finger antiviral protein recruits the RNA processing exosome to degrade the target mRNA. *Proc. Natl. Acad. Sci. USA* **2007**, *104*, 151–156. [CrossRef] [PubMed]

155. Zhu, Y.; Chen, G.; Lv, F.; Wang, X.; Ji, X.; Xu, Y.; Sun, J.; Wu, L.; Zheng, Y.-T.; Gao, G. Zinc-finger antiviral protein inhibits HIV-1 infection by selectively targeting multiply spliced viral mRNAs for degradation. *Proc. Natl. Acad. Sci. USA* **2011**, *108*, 15834–15839. [CrossRef] [PubMed]

156. Liu, L.; Chen, G.; Ji, X.; Gao, G. ZAP is a CRM1-dependent nucleocytoplasmic shuttling protein. *Biochem. Biophys. Res. Commun.* **2004**, *321*, 517–523. [CrossRef] [PubMed]

157. Molleston, J.M.; Sabin, L.R.; Moy, R.H.; Menghani, S.V.; Rausch, K.; Gordesky-Gold, B.; Hopkins, K.C.; Zhou, R.; Jensen, T.H.; Wilusz, J.E.; et al. A conserved virus-induced cytoplasmic TRAMP-like complex recruits the exosome to target viral RNA for degradation. *Genes Dev.* **2016**, *30*, 1658–1670. [CrossRef] [PubMed]

viruses

MDPI

Review

Use of Cellular Decapping Activators by Positive-Strand RNA Viruses

Jennifer Jungfleisch, Bernat Blasco-Moreno and Juana Díez *

Molecular Virology Laboratory, Department of Experimental and Health Sciences, Universitat Pompeu Fabra, Barcelona 08003, Spain; Jennifer.Jungfleisch@upf.edu (J.J.); Bernat.Blasco@upf.edu (B.B.-M.)
* Correspondence: juana.diez@upf.edu; Tel.: +34-933-160862

Academic Editor: J. Robert Hogg
Received: 24 October 2016; Accepted: 19 December 2016; Published: 21 December 2016

Abstract: Positive-strand RNA viruses have evolved multiple strategies to not only circumvent the hostile decay machinery but to trick it into being a priceless collaborator supporting viral RNA translation and replication. In this review, we describe the versatile interaction of positive-strand RNA viruses and the 5'-3' mRNA decay machinery with a focus on the viral subversion of decapping activators. This highly conserved viral trickery is exemplified with the plant Brome mosaic virus, the animal Flock house virus and the human hepatitis C virus.

Keywords: mRNA decay; positive strand RNA viruses; virus–host interactions

1. Introduction

Viruses maintain a constant duel with their infected host cells. Not only do they evolve strategies to circumvent deleterious cellular responses, but they also take advantage of the rich pools of host factors as controllable resources. Hijacking these resources is essential for the expansion of viruses as their gene-coding capacity is limited. One exquisite example of such viral manipulation is the use of the cellular mRNA decay machinery by a group of positive-strand RNA ((+)RNA) viruses. They use different strategies to turn the mRNA decay proteins into allies that support their replication and expansion.

The (+)RNA virus group includes numerous plant, animal and human pathogens such as the hepatitis C virus (HCV) and the emerging mosquito-borne Zika virus (ZIKV), Dengue virus (DENV), West Nile virus (WNV) and Chikungunya virus (CHIKV). Despite their diversity in terms of genome organization or virion morphology, the replication cycle of (+)RNA viruses is highly conserved [1]. Upon entering the target cell and gaining access to the cytoplasm, their single-stranded RNA genomes act like mRNAs and are directly translated by the host translation machinery to express the viral proteins. Once viral proteins accumulate, translation is repressed and the viral genomes are specifically recruited from the cellular translation machinery into membrane-bound viral replication complexes, where they act as templates for replication. Thus, to ensure productive infection, (+)RNA genomes must express enough viral protein to initiate the replication process and must keep their 5' and 3' ends intact to synthesize functional copies of the viral RNA. Hence, it is no surprise that for (+)RNA viruses to expand, the host mRNA decay machinery must be tricked (reviewed in [2]).

Cytoplasmic mRNA decay occurs via two major pathways—the deadenylation-dependent 5'-3' decay and the exonucleolytic 3'-5' decay pathways—that are conserved in all eukaryotes (Figure 1) [3]. In both pathways, the 3' poly(A) tail protects the mRNAs from degradation. Consequently, mRNAs are targeted to degradation only after the deadenylation complex Ccr4/Pop2/Not or Pan2/Pan3 [4,5] shortens the 3' poly(A)-tail. This leads to the opening of the closed-loop messenger ribonucleoprotein particle (mRNP) formed between the poly(A)-binding protein (PABP) and the cap-complex. Deadenylation is modulated by translation per se, by RNA binding proteins or by stress. Once deadenylation has occurred, mRNAs undergo degradation in the 5′-3′ direction, via decapping and subsequent degradation, or/and in the 3′-5′ direction, via the exosome complex. Besides these two major mRNA decay pathways, there are several specialized ones that primarily function in response to aberrancies in translation and are hence called mRNA quality control pathways (reviewed in [6]). These pathways are based on either deadenylation-independent decay [7], rapid 3′ to 5′ decay [8] or endonuclease cleavage [9] and include the nonsense-mediated decay (NMD), the no-go decay (NGD) and the non-stop decay (NSD) pathway.

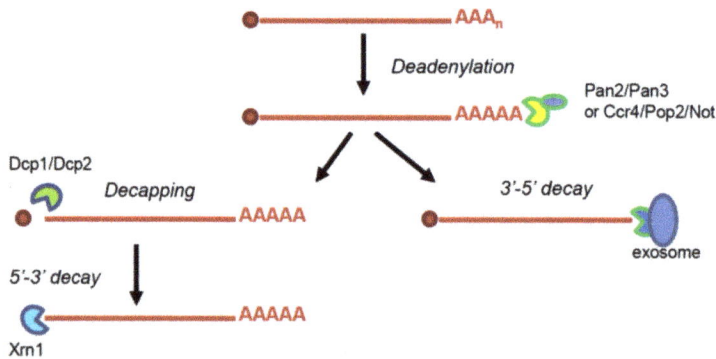

Figure 1. The two main mRNA decay pathways in the cytoplasm: The deadenylation-dependent 5′-3′ decay pathway and the exonucleolytic 3′-5′ decay pathway. Pan2/Pan3: PAB-dependent poly(A)-specific ribonuclease subunits; Pop: PGK promoter directed OverProduction; Not: Negative regulator of transcription subunit; Ccr4: C-C motif chemokine receptor 4; Dcp1/Dcp2: mRNA-decapping enzyme subunit 1/2; Xrn1: 5′-3′ exoribonuclease 1.

The deadenylation-dependent 5′-3′ decay pathway is the main cytoplasmic decay pathway (reviewed in [3]). mRNA-decapping enzyme subunit 2 (Dcp2) and 5′-3′ exoribonuclease 1 (Xrn1) are the two key enzymes in this pathway. Dcp2 cleaves the cap structure at the 5′ end of the mRNA, releasing a 7-methylguanosine diphosphate (m^7GDP) and a 5′ monophosphate mRNA. To be fully active, Dcp2 requires a conformational change mediated by the Dcp1 protein. The other enzyme is the 5′-3′ exonuclease Xrn1 that degrades the mRNA after decapping. In addition, there are other factors, named decapping activators, that assist and enhance the efficacy of the pathway. They include Sm-like proteins 1–7 (Lsm1–7), DNA topoisomerase 2-associated protein (Pat1), DExD/H-box ATP-dependent RNA helicase 1 (Dhh1), enhancer of mRNA-decapping protein 1–3 (Edc1–3) and the Suppressor of Clathrin Deficiency (Scd6). From these, Lsm1–7, Pat1 and Dhh1 are the most characterized ones. The Lsm1–7 ring is constituted by seven Lsm proteins that belong to the conserved Sm family of proteins and acts as an RNA chaperone facilitating a variety of RNA-RNA and RNA-protein interactions [10]. Pat1 functions as a scaffold protein, allowing the sequential binding of decay factors on mRNPs that eventually leads to degradation [11]. Lsm1–7 and Pat1 are co-purified from yeast extracts as a complex [12]. Finally, Dhh1 belongs to the family of DEAD box helicases characterized by acting as RNA chaperones as well. In the 5′-3′ deadenylation-dependent decay pathway, degradation occurs in three-steps. First, translation initiation is inhibited, as mRNA translation initiation and mRNA decay are connected processes in dynamic competition. This is achieved by the concerted action of the deadenylation complex and the decapping activators Dhh1, Lsm1–7, Pat1 and Scd6. Deadenylation leaves the 5′ cap structure accessible for the decapping complex Dcp1/Dcp2 while Dhh1, Pat1 and Scd6 inhibit translation initiation [13–15]. Dhh1 also hinders translation elongation by slowing down the ribosomes [16]. Although repression of translation initiation is required for decapping, it does not inevitably lead the mRNA to decapping as some mRNAs remain in a translationally-repressed state. Such translationally-repressed mRNAs can be stored in processing bodies (P-bodies), non-membranous dynamic cytoplasmic foci, and go back to translation or be further processed for degradation [17,18]. Second, the 5′ cap is removed by the Dcp1/Dcp2 decapping complex. The activity of the complex is accelerated by the decapping activators Lsm1–7, Edc1–3 and Pat1 [14,19–21]. Third, the exonuclease Xrn1 now has access to the uncapped mRNA and degrades it in the 5′-3′ direction.

The (+)RNA viruses have developed a myriad of strategies to shield their RNA from degradation by Xrn1, often by directly suppressing or degrading the cellular decay machinery. For example, picornaviruses use an aggressive mechanism to combat decay by inducing the rapid degradation of Xrn1 and Dcp1 [22,23]. WNV and DENV use Xrn1 to specifically generate sub-genomic flavivirus RNA (sfRNA) from the genomic RNA (gRNA) [24]. Degradation of gRNA by Xrn1 is stopped by a highly conserved RNA structure at the beginning of the 3′ untranslated region (UTR). The generated sfRNA plays essential roles in viral replication and pathogenesis in human hosts [24] and in mosquitoes as it inhibits the RNA interference (RNAi) response [25] and determines the infection and transmission rates [26]. Interestingly, the generated sfRNA displays an additional role. It binds to and inhibits Xrn1, hence, the endogenous mRNA turnover is altered [27–29]. This deregulated host mRNA stability is directly related to sfRNA expression and plays an important role in pathogenesis. As found for WNV and DENV, both the HCV and the Bovine viral diarrhea virus (BVDV) contain regions that stall and repress the enzymatic activity of Xrn1 [30]. However, in these two viruses, the regions are located in the 5′ UTRs. Intriguingly, other (+)RNA viruses, rather than avoiding or using the degradation activity of the 5′-3′ deadenylation-dependent decay machinery, redirect it to other functions. This review describes the strategies of the Brome mosaic virus (BMV), the Flock house virus (FHV) and HCV as representative plant, animal and human (+)RNA viruses that subvert the cellular decapping machinery to promote translation and replication of their viral RNA genomes.

2. The Brome Mosaic Virus Converts Enemies into Collaborators in Order to Promote Viral RNA Translation and Replication

A fruitful model system to study (+)RNA virus–host interactions is the replication of the plant BMV in *Saccharomyces cerevisiae* (reviewed in [31,32]). The BMV genome consists of three RNAs that are capped at their 5′ end, and at their 3′ end carry a conserved tRNA-like structure (TLS) instead of a poly(A) tail. Both UTRs contain overlapping sequences that control translation and the initiation of negative-strand synthesis (reviewed in [33]). RNA1 and RNA2 encode helicase 1a and polymerase 2a, respectively. The helicase 1a protein is the only viral protein required to recruit the BMV genome from the cellular translation machinery to the viral replication complex. RNA3 encodes the movement protein 3a and through a sub-genomic RNA generated during replication, the coat protein. Both proteins are required for the systemic infection of plants but not for viral replication.

Studies with the BMV/yeast model system led to the identification and characterization of hundreds of host factors required for different steps in the BMV life cycle [34,35]. Three unexpected factors were the decapping activators Lsm1–7, Pat1 and Dhh1. Depletion of the Lsm1–7, Pat1 or Dhh1 proteins dramatically reduced both BMV RNA translation and recruitment from translation to replication of the BMV RNA genomes [36–41]. Other components of the decay machinery are not required for these functions, indicating that BMV specifically subverts a selected group of decay factors [39]. The role of the Lsm1–7/Pat1 complex and the RNA helicase Dhh1 in translation has been thoroughly characterized. Both the Lsm1–7/Pat1 complex integrity and its intrinsic RNA-binding activity are required for translation of BMV RNAs [35]. The Lsm1–7/Pat1 complex directly interacts with sequences in both BMV RNA UTRs and with two internal A-rich single-stranded regions located in one of the BMV RNAs [34,35]. These sequences include well-characterized RNA motifs that control BMV RNA translation and replication. In turn, the helicase Dhh1 directly interacts with BMV RNA 3′ UTRs but not with the 5′ UTR. However, Dhh1 interacts with the translation initiation factors eIF4E, eIF4A and eIF4G located at the 5′ UTRs [42]. In addition, Dhh1 was found to bind sequences within the open reading frame (ORF) of BMV RNA2 (Figure 2). Importantly, Lsm1–7/Pat1 and Dhh1 targeted sequences are linked to the dependence on Lsm1–7/Pat1 and Dhh1 for translation, suggesting that their intrinsic RNA binding characteristics determine their function [37,39,42]. Recent exciting results indicate that Dhh1 also promotes translation of a specific set of cellular mRNAs encoding membrane and secreted proteins [42]. Viral and cellular Dhh1-dependent mRNA share some common key features. First, they contain long and highly structured 5′ UTRs and ORFs, including a region located close to the starting AUG. Second, they are directly bound by Dhh1 with a specific binding distribution. Third, they are likely activated by Dhh1 at the translation initiation step. Last, they encode proteins that localize in membranes. Whether Lsm1–7/Pat1 may exert a similar function on cellular mRNAs remains unknown.

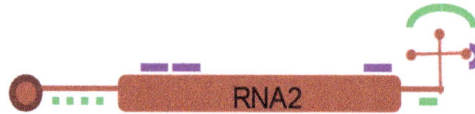

Figure 2. Binding pattern for DExD/H-box ATP-dependent RNA helicase 1 (Dhh1) and the Lsm1–7/Pat1 complex to brome mosaic virus (BMV) RNA2. Dhh1 binds to three sites in the open reading frame (ORF) and to the tRNA-like structure (TLS) in the 3′ untranslated region (UTR) (purple color). The Lsm1–7/Pat1 complex binds to both the TLS and the non-tRNA-like structure in the 3′ UTR and also with lower affinity to the 5′ UTR (green color).

The mechanisms by which the Lsm1–7/Pat1 complex promotes translation and recruitment of BMV RNAs are different. Mutations in the *LSM1* gene, a key component of the Lsm1–7 ring, affect differentially BMV RNA translation and recruitment [35]. Importantly, the Lsm1–7 Pat1 complex interacts in a RNAse-resistent manner with the BMV 1a, the solely viral protein required for recruiment. In line with this, the RNA-biding activity of the Lsm1–7/Pat1 complex is not requried for its function in recruitment [38]. The molecular mechanisms by which Dhh1 promotes BMV RNA recruitment remain uncharacterized. Taken together, the decapping activators Lsm1–7, Pat1 and Dhh1 bind specifically to the BMV RNA genome, promoting its translation and replication rather than its decay. Current results support a model (Figure 3) in which Lsm1–7, Pat1 and Dhh1 bind to *cis*-elements in the viral RNA, thereby remodeling mRNA secondary structures and promoting its circularization and translation. Since poly(A)-tails in cellular mRNAs mediate 5′-3′ circularization via the binding of the poly(A) binding protein, binding of Lsm1–7, Pat1 and Dhh1 to the 5′ and 3′ UTRs structure and to initiation factors at the 5′ UTR would establish such circularization in BMV RNAs. During recruitment, circularization would be disrupted as the Lsm1–7/Pat1 complex would now bind to the viral 1a protein, driving the viral RNA from translation to replication.

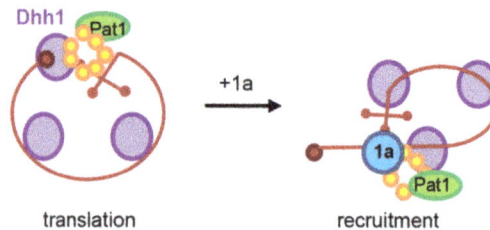

Figure 3. Model of the Lsm1–7/Pat1 complex and Dhh1 function in viral RNA translation and recruitment.

3. The Flock House Virus Subverts Features of Decapping Proteins to Control the Genomic to Sub-Genomic Viral RNA Ratio

The FHV (reviewed in [43]) is an insect pathogen that can replicate in a wide variety of hosts, including *Drosophila*, *Caenorhabditis elegans*, plants, mammals and *Saccharomyces cerevisiae*. Accordingly, the host factors hijacked by FHV to replicate are highly conserved [44–47]. The simplicity of the FHV genome, combined with the advantages of yeast genetics make the FHV-yeast system another excellent model system to study basic aspects of (+)RNA biology, including virus–host interactions. Interestingly, three of the highly conserved host factors subverted by FHV are the decapping activators Lsm1–7, Pat1 and Dhh1 [48].

The FHV bipartite genome consists of two capped but non-polyadenylated RNA segments. RNA1 encodes protein A, the only FHV protein required for replication, and RNA2 encodes the capsid precursor. During replication, RNA1 also produces sub-genomic RNA3 that encodes protein B1, of unknown function, and protein B2, required to suppress RNA silencing in infected hosts. RNA3 corresponds to the 3′ end of RNA1 and it is synthesized during RNA1 replication. In addition to its coding function, RNA3 coordinates the production of appropriate levels of RNA1 and RNA2 [49–51]. This latter activity is essential for proper and timely expression of the different viral proteins throughout the different stages of the viral life cycle. Interestingly, depletion of Lsm1–7, Pat1 or Dhh1 disrupts this activity and alters the RNA1/RNA3 ratio [48].

Lsm1–7, Pat1 and Dhh1 control RNA3 synthesis [48]. Different RNP rearrangements of the genomic RNA1 are necessary for the viral polymerase to synthesize a complete copy of RNA1 or a partial one, RNA3 [50]. RNA3 synthesis requires a long-distance base pairing interaction between *cis*-elements in RNA1 [50]. These base pairing interactions stop the polymerase prematurely and lead to the synthesis of RNA3 instead of RNA1. As Dhh1 remodels RNP compositions in an ATPase-dependent manner [52] and RNA3 synthesis requires the ATPase activity of the Dhh1 helicase [48], Lsm1–7, Pat1 and Dhh1 have been proposed to regulate the key viral RNP transitions required to maintain the balance between the alternative FHV RNA1 conformations controlling RNA3 synthesis [48]. Interestingly, and as found for BMV, Lsm1–7, Pat1 and Dhh1 interact not only with the RNA genome but also with the viral polymerase [48].

4. Subverting Decapping Activators Is Conserved in Human (+)RNA Viruses

Since the 5′-3′ decay pathway is strongly conserved from yeast to humans, human (+)RNA viruses might as well subvert it to favor viral replication. Indeed, the human counterparts of Lsm1–7, Pat1 and Dhh1—namely Lsm1–7, PatL1 and DEAD-Box Helicase 6 (DDX6)—are required for the expansion of the flaviviruses HCV, DENV and WNV. Depletion of Lsm1–7, PatL1 or DDX6 directly and specifically inhibits HCV RNA translation and replication [53,54]. Moreover, in vitro binding assays demonstrated direct interactions of human Lsm1–7 complexes with essential translation/replication regulatory sequences in the 5′ and 3′ UTRs of the HCV RNA genome [53]. DDX6 also interacts with the HCV RNA genome and the core protein in HCV-infected cells [55]. Likewise, DDX6 and Lsm1 promote replication of the WNV and co-localize with the viral replication complex [56]. Moreover, DDX6 directly interacts with structured regulatory *cis*-sequences in the DENV RNA genome and it co-localizes with the DENV replication complex [57].

Although all the above described examples of (+)RNA viruses belong to the *Flaviviridae* family, the conserved use of decapping factors by (+)strand RNA viruses that infect plants (BMV), insects (FHV) and even bacteria (Qβ [58]) underlines the robustness of this strategy to regulate (+)RNA virus life cycles and suggests that it extends to other human viruses outside the *Flaviviridae* family. Depriving (+)RNA viruses of this highly conserved strategy by targeting these decay factors with drugs would therefore appear to be a promising strategy to generate broad-spectrum antiviral drugs. The fact that the individual, transient knock-down of Lsm1–7, PatL1 or DDX6 proteins in human cells is not toxic and that the respective yeast knockout strains are viable, stresses the feasibility of such an approach for the future.

What are the common features of the aforementioned activities of Lsm1–7, Pat1/PatL1 and Dhh1/DDX6 in (+)RNA viral life cycles? In all cases, the decapping activators interact with viral replication proteins, and/or with specific and structured regulatory *cis*-acting signals in the viral RNA genome. The role of these decapping proteins as catalyzers of mRNP transitions that direct cellular mRNAs from translation to decay suggests that they act similarly on the highly structured viral (+)RNA genomes, directing them to translation or replication.

5. (+)RNA Viruses Alter the Distribution of Decay Factors

Lsm1–7, PatL1 and DDX6 accumulate in P-bodies [59,60]. P-bodies are discrete and highly dynamic cytoplasmic mRNP granules found in eukaryotic cells under normal growth condition [59]. These structures contain translationally-repressed mRNAs together with multiple proteins from the 5′-3′ mRNA decay and silencing machineries [60]. Once in P-bodies, mRNAs can be either degraded or stored for future translation [17,61,62]. The components of P-bodies cycle rapidly in and out of these granules, indicating that there is a constant exchange with the cytoplasm where all these components are diffusely distributed [63–65]. Importantly, the formation of P-bodies requires most of its components [11,63,66–69]. Consequently, conditions that reduce the cytosolic concentration of P-body proteins, such as (+)RNA virus infection, disrupt the formation of P-bodies (Figure 4). For example, the picornaviruses poliovirus and Coxsackie virus disrupt P-bodies by degrading the P-body core components Xrn1 and Dcp1a [22,23], and HCV, WNV and DENV do so by preventing Lsm1–7, PatL1 and DDX6 from participating in P-body formation [70–72]. All these results were obtained in cell culture infection systems. Importantly, HCV inhibits P-body granule formation in human livers regardless of viral phenotype, inflammation grade or whether infection was recent or long established. Moreover, this alteration is reversed once HCV is eliminated by therapy. Therefore, there is a link between P-body alterations and pathogenic conditions [73].

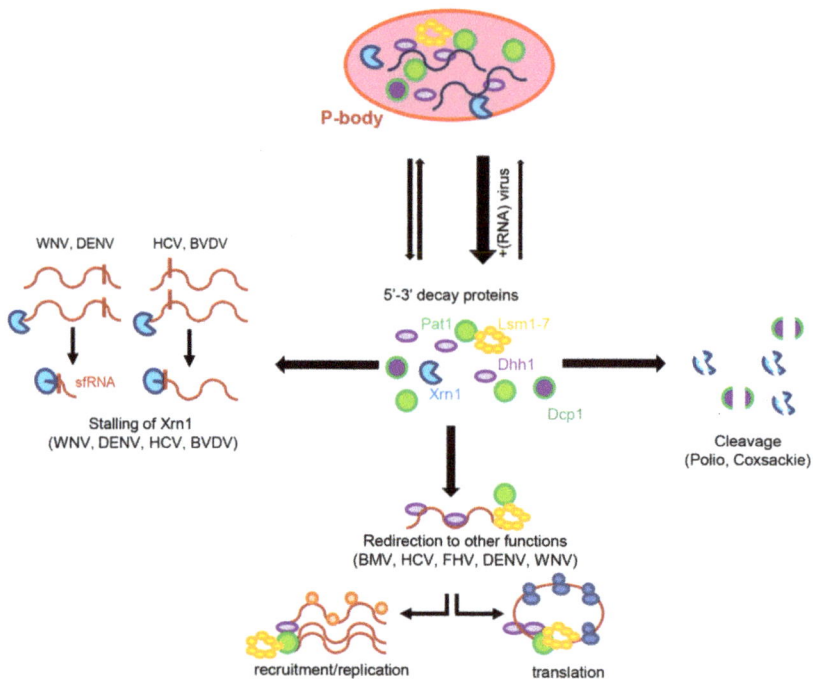

Figure 4. (+)RNA viruses and the cellular mRNA decay machinery. Multiple strategies have been developed by (+)RNA viruses to not only prevent the degradation of the viral RNA genomes but also, to subvert it to their benefit. P-bodies: processing bodies; WNV: West Nile virus; DENV: Dengue virus; HCV: hepatitis C virus; BVDV: bovine viral diarrhea virus; FHV: Flock house virus; BMV: brome mosaic virus.

The obvious question to ask is whether P-body disruption is required for viral infection or whether it is just a consequence. The data currently available suggest that it is a mere consequence, as P-body formation is not required for mRNA decay [74] and depletion of the P-body component Rap55 disrupts P-body formation but it does not affect HCV expansion [70]. Irrespective of the question whether P-body disruption is the consequence or the trigger of pathogenesis, it seems obvious that changing the equilibrium of granulated versus free P-body proteins, proteins that control decay and silencing, would alter the transcriptome and translatome and consequently gene expression. These alterations might be linked to different viral pathologies. Accordingly, to further study the interactions of (+)RNA viruses with P-bodies represents an interesting field that may open many opportunities in terms of therapeutic strategies.

6. Concluding Remarks

Viruses are masters in converting hostile conditions into a paradise for their own replication. The sequestering of the cellular decay machinery by (+)RNA viruses is a remarkable example of this that goes beyond using the decay components themselves towards altering the whole transcriptional/translational landscape of the host. The strong conservation of this viral strategy across species pinpoints a weak spot that can be exploited for the development of broad-spectrum antiviral drugs. Yet, some essential questions remain open. For example, how are viral infection, the host transcriptome/translatome and pathogenesis linked? How do viruses regulate host transcription to favor their own replication? The basic nature of these questions highlights that scientists in this area still have a long path to go.

Acknowledgments: This work was supported by the Spanish Ministry of Economy and Competitiveness through grant BFU 2013-44629-R and the "Maria de Maeztu" Programme for Units of Excellence in R&D (MDM-2014-0370).

Conflicts of Interest: The authors declare no conflict of interest.

References

1. Ahlquist, P. Parallels among positive-strand RNA viruses, reverse-transcribing viruses and double-stranded RNA viruses. *Nat. Rev. Microbiol.* **2006**, *4*, 371–382. [CrossRef] [PubMed]
2. Moon, S.L.; Wilusz, J. Cytoplasmic viruses: Rage against the (cellular RNA decay) machine. *PLoS Pathog.* **2013**, *9*, e1003762. [CrossRef] [PubMed]
3. Parker, R. RNA degradation in saccharomyces cerevisae. *Genetics* **2012**, *191*, 671–702. [CrossRef] [PubMed]
4. Daugeron, M.C.; Mauxion, F.; Seraphin, B. The yeast pop2 gene encodes a nuclease involved in mRNA deadenylation. *Nucleic Acids Res.* **2001**, *29*, 2448–2455. [CrossRef] [PubMed]
5. Tucker, M.; Staples, R.R.; Valencia-Sanchez, M.A.; Muhlrad, D.; Parker, R. Ccr4p is the catalytic subunit of a ccr4p/pop2p/notp mRNA deadenylase complex in saccharomyces cerevisiae. *EMBO J.* **2002**, *21*, 1427–1436. [CrossRef] [PubMed]
6. Shoemaker, C.J.; Green, R. Translation drives mRNA quality control. *Nat. Struct. Mol. Biol.* **2012**, *19*, 594–601. [CrossRef] [PubMed]
7. Muhlrad, D.; Parker, R. Premature translational termination triggers mRNA decapping. *Nature* **1994**, *370*, 578–581. [CrossRef] [PubMed]
8. van Hoof, A.; Frischmeyer, P.A.; Dietz, H.C.; Parker, R. Exosome-mediated recognition and degradation of mRNAs lacking a termination codon. *Science* **2002**, *295*, 2262–2264. [CrossRef] [PubMed]
9. Doma, M.K.; Parker, R. Endonucleolytic cleavage of eukaryotic mRNAs with stalls in translation elongation. *Nature* **2006**, *440*, 561–564. [CrossRef] [PubMed]
10. Wilusz, C.J.; Wilusz, J. Eukaryotic lsm proteins: Lessons from bacteria. *Nat. Struct. Mol. Biol.* **2005**, *12*, 1031–1036. [CrossRef] [PubMed]
11. Marnef, A.; Standart, N. Pat1 proteins: A life in translation, translation repression and mRNA decay. *Biochem. Soc. Trans.* **2010**, *38*, 1602–1607. [CrossRef] [PubMed]
12. Tharun, S. Purification and analysis of the decapping activator lsm1p-7p-pat1p complex from yeast. *Methods Enzymol.* **2008**, *448*, 41–55. [PubMed]

13. Coller, J.; Parker, R. General translational repression by activators of mRNA decapping. *Cell* **2005**, *122*, 875–886. [CrossRef] [PubMed]
14. Nissan, T.; Rajyaguru, P.; She, M.; Song, H.; Parker, R. Decapping activators in saccharomyces cerevisiae act by multiple mechanisms. *Mol. Cell* **2010**, *39*, 773–783. [CrossRef] [PubMed]
15. Rajyaguru, P.; She, M.; Parker, R. Scd6 targets eif4g to repress translation: Rgg motif proteins as a class of eif4g-binding proteins. *Mol. Cell* **2012**, *45*, 244–254. [CrossRef] [PubMed]
16. Sweet, T.; Kovalak, C.; Coller, J. The dead-box protein dhh1 promotes decapping by slowing ribosome movement. *PLoS Biol.* **2012**, *10*, e1001342. [CrossRef] [PubMed]
17. Aizer, A.; Kalo, A.; Kafri, P.; Shraga, A.; Ben-Yishay, R.; Jacob, A.; Kinor, N.; Shav-Tal, Y. Quantifying mRNA targeting to p-bodies in living human cells reveals their dual role in mRNA decay and storage. *J. Cell Sci.* **2014**, *127*, 4443–4456. [CrossRef] [PubMed]
18. Arribere, J.A.; Doudna, J.A.; Gilbert, W.V. Reconsidering movement of eukaryotic mRNAs between polysomes and p bodies. *Mol. Cell* **2011**, *44*, 745–758. [CrossRef] [PubMed]
19. Dunckley, T.; Tucker, M.; Parker, R. Two related proteins, edc1p and edc2p, stimulate mRNA decapping in saccharomyces cerevisiae. *Genetics* **2001**, *157*, 27–37. [PubMed]
20. Harigaya, Y.; Jones, B.N.; Muhlrad, D.; Gross, J.D.; Parker, R. Identification and analysis of the interaction between edc3 and dcp2 in saccharomyces cerevisiae. *Mol. Cell Biol.* **2010**, *30*, 1446–1456. [CrossRef] [PubMed]
21. Schwartz, D.; Decker, C.J.; Parker, R. The enhancer of decapping proteins, edc1p and edc2p, bind RNA and stimulate the activity of the decapping enzyme. *RNA* **2003**, *9*, 239–251. [CrossRef] [PubMed]
22. Dougherty, J.D.; White, J.P.; Lloyd, R.E. Poliovirus-mediated disruption of cytoplasmic processing bodies. *J. Virol.* **2011**, *85*, 64–75. [CrossRef] [PubMed]
23. Rozovics, J.M.; Chase, A.J.; Cathcart, A.L.; Chou, W.; Gershon, P.D.; Palusa, S.; Wilusz, J.; Semler, B.L. PicoRNAvirus modification of a host mRNA decay protein. *MBio* **2012**, *3*, e00431-12. [CrossRef] [PubMed]
24. Pijlman, G.P.; Funk, A.; Kondratieva, N.; Leung, J.; Torres, S.; van der Aa, L.; Liu, W.J.; Palmenberg, A.C.; Shi, P.Y.; Hall, R.A.; et al. A highly structured, nuclease-resistant, noncoding RNA produced by flaviviruses is required for pathogenicity. *Cell Host Microbe* **2008**, *4*, 579–591. [CrossRef] [PubMed]
25. Schnettler, E.; Sterken, M.G.; Leung, J.Y.; Metz, S.W.; Geertsema, C.; Goldbach, R.W.; Vlak, J.M.; Kohl, A.; Khromykh, A.A.; Pijlman, G.P. Noncoding flavivirus RNA displays RNA interference suppressor activity in insect and mammalian cells. *J. Virol.* **2012**, *86*, 13486–13500. [CrossRef] [PubMed]
26. Goertz, G.P.; Fros, J.J.; Miesen, P.; Vogels, C.B.; van der Bent, M.L.; Geertsema, C.; Koenraadt, C.J.; van Rij, R.P.; van Oers, M.M.; Pijlman, G.P. Noncoding subgenomic flavivirus RNA is processed by the mosquito RNA interference machinery and determines west nile virus transmission by culex pipiens mosquitoes. *J. Virol.* **2016**, *90*, 10145–10159. [CrossRef] [PubMed]
27. Chapman, E.G.; Costantino, D.A.; Rabe, J.L.; Moon, S.L.; Wilusz, J.; Nix, J.C.; Kieft, J.S. The structural basis of pathogenic subgenomic flavivirus RNA (sfRNA) production. *Science* **2014**, *344*, 307–310. [CrossRef] [PubMed]
28. Chapman, E.G.; Moon, S.L.; Wilusz, J.; Kieft, J.S. RNA structures that resist degradation by xrn1 produce a pathogenic dengue virus RNA. *Elife* **2014**, *3*, e01892. [CrossRef] [PubMed]
29. Moon, S.L.; Anderson, J.R.; Kumagai, Y.; Wilusz, C.J.; Akira, S.; Khromykh, A.A.; Wilusz, J. A noncoding RNA produced by arthropod-borne flaviviruses inhibits the cellular exoribonuclease xrn1 and alters host mRNA stability. *RNA* **2012**, *18*, 2029–2040. [CrossRef] [PubMed]
30. Moon, S.L.; Blackinton, J.G.; Anderson, J.R.; Dozier, M.K.; Dodd, B.J.; Keene, J.D.; Wilusz, C.J.; Bradrick, S.S.; Wilusz, J. Xrn1 stalling in the 5′ utr of hepatitis c virus and bovine viral diarrhea virus is associated with dysregulated host mRNA stability. *PLoS Pathog.* **2015**, *11*, e1004708. [CrossRef] [PubMed]
31. Galao, R.P.; Scheller, N.; Alves-Rodrigues, I.; Breinig, T.; Meyerhans, A.; Diez, J. Saccharomyces cerevisiae: A versatile eukaryotic system in virology. *Microb. Cell Fact.* **2007**, *6*, 32. [CrossRef] [PubMed]
32. Nagy, P.D. Yeast as a model host to explore plant virus-host interactions. *Annu. Rev. Phytopathol.* **2008**, *46*, 217–242. [CrossRef] [PubMed]
33. Noueiry, A.O.; Ahlquist, P. Brome mosaic virus RNA replication: Revealing the role of the host in RNA virus replication. *Annu. Rev. Phytopathol.* **2003**, *41*, 77–98. [CrossRef] [PubMed]
34. Gancarz, B.L.; Hao, L.; He, Q.; Newton, M.A.; Ahlquist, P. Systematic identification of novel, essential host genes affecting bromovirus RNA replication. *PLoS ONE* **2011**, *6*, e23988. [CrossRef] [PubMed]

35. Kushner, D.B.; Lindenbach, B.D.; Grdzelishvili, V.Z.; Noueiry, A.O.; Paul, S.M.; Ahlquist, P. Systematic, genome-wide identification of host genes affecting replication of a positive-strand RNA virus. *Proc. Natl. Acad. Sci. USA* **2003**, *100*, 15764–15769. [CrossRef] [PubMed]

36. Alves-Rodrigues, I.; Mas, A.; Diez, J. Xenopus xp54 and human rck/p54 helicases functionally replace yeast dhh1p in brome mosaic virus RNA replication. *J. Virol.* **2007**, *81*, 4378–4380. [CrossRef] [PubMed]

37. Galao, R.P.; Chari, A.; Alves-Rodrigues, I.; Lobao, D.; Mas, A.; Kambach, C.; Fischer, U.; Diez, J. Lsm1–7 complexes bind to specific sites in viral RNA genomes and regulate their translation and replication. *RNA* **2010**, *16*, 817–827. [CrossRef] [PubMed]

38. Jungfleisch, J.; Chowdhury, A.; Alves-Rodrigues, I.; Tharun, S.; Diez, J. The lsm1–7-pat1 complex promotes viral RNA translation and replication by differential mechanisms. *RNA* **2015**, *21*, 1469–1479. [CrossRef] [PubMed]

39. Mas, A.; Alves-Rodrigues, I.; Noueiry, A.; Ahlquist, P.; Diez, J. Host deadenylation-dependent mRNA decapping factors are required for a key step in brome mosaic virus RNA replication. *J. Virol.* **2006**, *80*, 246–251. [CrossRef] [PubMed]

40. Diez, J.; Ishikawa, M.; Kaido, M.; Ahlquist, P. Identification and characterization of a host protein required for efficient template selection in viral RNA replication. *Proc. Natl. Acad. Sci. USA* **2000**, *97*, 3913–3918. [CrossRef] [PubMed]

41. Noueiry, A.O.; Diez, J.; Falk, S.P.; Chen, J.; Ahlquist, P. Yeast lsm1p–7p/pat1p deadenylation-dependent mRNA-decapping factors are required for brome mosaic virus genomic RNA translation. *Mol. Cell Biol.* **2003**, *23*, 4094–4106. [CrossRef] [PubMed]

42. Jungfleisch, J.; Nedialkova, D.D.; Dotu, I.; Sloan, K.E.; Martinez-Bosch, N.; Bruning, L.; Raineri, E.; Navarro, P.; Bohnsack, M.T.; Leidel, S.A.; et al. A novel translational control mechanism involving RNA structures within coding sequences. *Genome Res.* **2016**. [CrossRef] [PubMed]

43. Venter, P.A.; Schneemann, A. Recent insights into the biology and biomedical applications of flock house virus. *Cell. Mol. Life Sci.* **2008**, *65*, 2675–2687. [CrossRef] [PubMed]

44. Dasgupta, R.; Garcia, B.H., 2nd; Goodman, R.M. Systemic spread of an RNA insect virus in plants expressing plant viral movement protein genes. *Proc. Natl. Acad. Sci. USA* **2001**, *98*, 4910–4915. [CrossRef] [PubMed]

45. Johnson, K.L.; Ball, L.A. Replication of flock house virus RNAs from primary transcripts made in cells by RNA polymerase II. *J. Virol.* **1997**, *71*, 3323–3327. [PubMed]

46. Lu, R.; Maduro, M.; Li, F.; Li, H.W.; Broitman-Maduro, G.; Li, W.X.; Ding, S.W. Animal virus replication and RNAi-mediated antiviral silencing in caenorhabditis elegans. *Nature* **2005**, *436*, 1040–1043. [CrossRef] [PubMed]

47. Price, B.D.; Rueckert, R.R.; Ahlquist, P. Complete replication of an animal virus and maintenance of expression vectors derived from it in saccharomyces cerevisiae. *Proc. Natl. Acad. Sci. USA* **1996**, *93*, 9465–9470. [CrossRef] [PubMed]

48. Gimenez-Barcons, M.; Alves-Rodrigues, I.; Jungfleisch, J.; Van Wynsberghe, P.M.; Ahlquist, P.; Diez, J. The cellular decapping activators lsm1, pat1, and dhh1 control the ratio of subgenomic to genomic flock house virus RNAs. *J. Virol.* **2013**, *87*, 6192–6200. [CrossRef] [PubMed]

49. Eckerle, L.D.; Ball, L.A. Replication of the RNA segments of a bipartite viral genome is coordinated by a transactivating subgenomic RNA. *Virology* **2002**, *296*, 165–176. [CrossRef] [PubMed]

50. Lindenbach, B.D.; Sgro, J.Y.; Ahlquist, P. Long-distance base pairing in flock house virus RNA1 regulates subgenomic RNA3 synthesis and RNA2 replication. *J. Virol.* **2002**, *76*, 3905–3919. [CrossRef] [PubMed]

51. Zhong, W.; Rueckert, R.R. Flock house virus: Down-regulation of subgenomic RNA3 synthesis does not involve coat protein and is targeted to synthesis of its positive strand. *J. Virol.* **1993**, *67*, 2716–2722. [PubMed]

52. Weston, A.; Sommerville, J. Xp54 and related (ddx6-like) RNA helicases: Roles in messenger rnp assembly, translation regulation and RNA degradation. *Nucl. Acids Res.* **2006**, *34*, 3082–3094. [CrossRef] [PubMed]

53. Scheller, N.; Mina, L.B.; Galao, R.P.; Chari, A.; Gimenez-Barcons, M.; Noueiry, A.; Fischer, U.; Meyerhans, A.; Diez, J. Translation and replication of hepatitis c virus genomic RNA depends on ancient cellular proteins that control mRNA fates. *Proc. Natl. Acad. Sci. USA* **2009**, *106*, 13517–13522. [CrossRef] [PubMed]

54. Huys, A.; Thibault, P.A.; Wilson, J.A. Modulation of hepatitis c virus RNA accumulation and translation by ddx6 and mir-122 are mediated by separate mechanisms. *PLoS ONE* **2013**, *8*, e67437. [CrossRef] [PubMed]

55. Jangra, R.K.; Yi, M.; Lemon, S.M. Ddx6 (rck/p54) is required for efficient hepatitis c virus replication but not for internal ribosome entry site-directed translation. *J. Virol.* **2010**, *84*, 6810–6824. [CrossRef] [PubMed]

56. Chahar, H.S.; Chen, S.; Manjunath, N. P-body components lsm1, gw182, ddx3, ddx6 and xrn1 are recruited to wnv replication sites and positively regulate viral replication. *Virology* **2013**, *436*, 1–7. [CrossRef] [PubMed]

57. Ward, A.M.; Bidet, K.; Yinglin, A.; Ler, S.G.; Hogue, K.; Blackstock, W.; Gunaratne, J.; Garcia-Blanco, M.A. Quantitative mass spectrometry of denv-2 RNA-interacting proteins reveals that the dead-box RNA helicase ddx6 binds the db1 and db2 3' utr structures. *RNA Biol.* **2011**, *8*, 1173–1186. [CrossRef] [PubMed]

58. Schuppli, D.; Miranda, G.; Tsui, H.C.; Winkler, M.E.; Sogo, J.M.; Weber, H. Altered 3'-terminal RNA structure in phage qbeta adapted to host factor-less escherichia coli. *Proc. Natl. Acad. Sci. USA* **1997**, *94*, 10239–10242. [PubMed]

59. Buchan, J.R. Mrnp granules: Assembly, function, and connections with disease. *RNA Biol.* **2014**, 11. [CrossRef] [PubMed]

60. Jain, S.; Parker, R. The discovery and analysis of p bodies. *Adv. Exp. Med. Biol.* **2013**, *768*, 23–43. [PubMed]

61. Cougot, N.; Babajko, S.; Seraphin, B. Cytoplasmic foci are sites of mRNA decay in human cells. *J. Cell Biol.* **2004**, *165*, 31–40. [CrossRef] [PubMed]

62. Brengues, M.; Teixeira, D.; Parker, R. Movement of eukaryotic mRNAs between polysomes and cytoplasmic processing bodies. *Science* **2005**, *310*, 486–489. [CrossRef] [PubMed]

63. Andrei, M.A.; Ingelfinger, D.; Heintzmann, R.; Achsel, T.; Rivera-Pomar, R.; Luhrmann, R. A role for eif4e and eif4e-transporter in targeting mrnps to mammalian processing bodies. *RNA* **2005**, *11*, 717–727. [CrossRef] [PubMed]

64. Aizer, A.; Brody, Y.; Ler, L.W.; Sonenberg, N.; Singer, R.H.; Shav-Tal, Y. The dynamics of mammalian p body transport, assembly, and disassembly in vivo. *Mol. Biol. Cell* **2008**, *19*, 4154–4166. [CrossRef] [PubMed]

65. Kedersha, N.; Stoecklin, G.; Ayodele, M.; Yacono, P.; Lykke-Andersen, J.; Fritzler, M.J.; Scheuner, D.; Kaufman, R.J.; Golan, D.E.; Anderson, P. Stress granules and processing bodies are dynamically linked sites of mrnp remodeling. *J. Cell Biol.* **2005**, *169*, 871–884. [CrossRef] [PubMed]

66. Yang, W.H.; Yu, J.H.; Gulick, T.; Bloch, K.D.; Bloch, D.B. RNA-associated protein 55 (rap55) localizes to mRNA processing bodies and stress granules. *RNA* **2006**, *12*, 547–554. [CrossRef] [PubMed]

67. Yu, J.H.; Yang, W.H.; Gulick, T.; Bloch, K.D.; Bloch, D.B. Ge-1 is a central component of the mammalian cytoplasmic mRNA processing body. *RNA* **2005**, *11*, 1795–1802. [CrossRef] [PubMed]

68. Ozgur, S.; Chekulaeva, M.; Stoecklin, G. Human pat1b connects deadenylation with mRNA decapping and controls the assembly of processing bodies. *Mol. Cell Biol.* **2010**, *30*, 4308–4323. [CrossRef] [PubMed]

69. Ayache, J.; Benard, M.; Ernoult-Lange, M.; Minshall, N.; Standart, N.; Kress, M.; Weil, D. P-body assembly requires ddx6 repression complexes rather than decay or ataxin2/2l complexes. *Mol. Biol. Cell* **2015**, *26*, 2579–2595. [CrossRef] [PubMed]

70. Perez-Vilaro, G.; Scheller, N.; Saludes, V.; Diez, J. Hepatitis c virus infection alters p-body composition but is independent of p-body granules. *J. Virol.* **2012**, *86*, 8740–8749. [CrossRef] [PubMed]

71. Emara, M.M.; Brinton, M.A. Interaction of tia-1/tiar with west nile and dengue virus products in infected cells interferes with stress granule formation and processing body assembly. *Proc. Natl. Acad. Sci. USA* **2007**, *104*, 9041–9046. [CrossRef] [PubMed]

72. Ariumi, Y.; Kuroki, M.; Kushima, Y.; Osugi, K.; Hijikata, M.; Maki, M.; Ikeda, M.; Kato, N. Hepatitis c virus hijacks p-body and stress granule components around lipid droplets. *J. Virol.* **2011**, *85*, 6882–6892. [CrossRef] [PubMed]

73. Perez-Vilaro, G.; Fernandez-Carrillo, C.; Mensa, L.; Miquel, R.; Sanjuan, X.; Forns, X.; Perez-del-Pulgar, S.; Diez, J. Hepatitis c virus infection inhibits p-body granule formation in human livers. *J. Hepatol.* **2015**, *62*, 785–790. [CrossRef] [PubMed]

74. Eulalio, A.; Behm-Ansmant, I.; Schweizer, D.; Izaurralde, E. P-body formation is a consequence, not the cause, of RNA-mediated gene silencing. *Mol. Cell Biol.* **2007**, *27*, 3970–3981. [CrossRef] [PubMed]

viruses

Review

Diverse Strategies Used by Picornaviruses to Escape Host RNA Decay Pathways

Wendy Ullmer and Bert L. Semler *

Department of Microbiology and Molecular Genetics, School of Medicine, University of California, Irvine, CA 92697, USA; wullmer@uci.edu
* Correspondence: blsemler@uci.edu; Tel.: +1-949-824-7573

Academic Editor: Karen Beemon
Received: 5 November 2016; Accepted: 9 December 2016; Published: 20 December 2016

Abstract: To successfully replicate, viruses protect their genomic material from degradation by the host cell. RNA viruses must contend with numerous destabilizing host cell processes including mRNA decay pathways and viral RNA (vRNA) degradation resulting from the antiviral response. Members of the *Picornaviridae* family of small RNA viruses have evolved numerous diverse strategies to evade RNA decay, including incorporation of stabilizing elements into vRNA and re-purposing host stability factors. Viral proteins are deployed to disrupt and inhibit components of the decay machinery and to redirect decay machinery to the advantage of the virus. This review summarizes documented interactions of picornaviruses with cellular RNA decay pathways and processes.

Keywords: picornavirus; *Picornaviridae*; poliovirus; coxsackievirus; human rhinovirus; RNA degradation; mRNA decay; RNA stability; RNase L; deadenylase

1. Introduction

Cytoplasmic RNA viruses encounter a myriad of host defense mechanisms that must be countered by a small arsenal of viral proteins. Preserving the stability and integrity of viral RNA (vRNA) is of fundamental importance to the virus to ensure successful generation of progeny virions. Throughout the virus replication cycle, vRNAs encounter multiple potentially destabilizing host cell pathways and processes, from regulated mRNA decay pathways to interferon (IFN)-induced vRNA decay. Members of the *Picornaviridae* family are small, positive-sense single-stranded RNA viruses that have evolved strategies to re-purpose, inhibit, or otherwise evade many components of the cellular RNA decay machinery.

As a family, picornaviruses are composed of at least 29 different genera which include many significant human and animal pathogens causing a range of illnesses and economic burden. The Enterovirus genus of picornaviruses includes the causative agents of paralytic poliomyelitis (poliovirus), hand, foot, and mouth disease (coxsackievirus (CVA) A16 and enterovirus (EV) 71), and the common cold (human rhinovirus (HRV)). Other severe symptoms of EV infection include meningitis, encephalitis, myocarditis, and pericarditis, which can arise from infection by subtypes of CVA, coxsackievirus B (CVB), EV, or echovirus. The lone member of the Hepatovirus genus, hepatitis A virus (HAV), infects the liver and, in rare cases, causes acute liver failure. The Cardiovirus genus includes encephalomyocarditis virus (EMCV), Theiler's murine encephalomyelitis virus (TMEV), and Saffold virus (SAFV). EMCV and TMEV are largely non-human pathogens that cause symptoms that include myocarditis, encephalitis, and, for EMCV specifically, reproductive failure in pigs. SAFV is a recently discovered human cardiovirus that does not yet have clearly defined pathological features but has been linked to acute flaccid paralysis, meningitis, and cerebellitis. Foot-and-mouth disease virus (FMDV), a member of the Aphthovirus genus, is one of the most economically important livestock viruses. Infection by

FMDV causes painful vesicles in the feet, mouth, and teats of cloven-hoofed animals, reducing their productivity and requiring significant eradication measures.

Picornaviruses induce extensive modification of cellular processes to complete their replication cycle. Upon attachment and release of genomic RNA into the cytoplasm of a host cell, viral proteins are translated by a cap-independent mechanism using an internal ribosome entry site (IRES) located in the 5′ non-coding region (5′ NCR) of the vRNA. Picornavirus genomic RNAs are linked to a small, virus-derived protein at their 5′ termini (VPg), possess a 3′ poly(A) tract, and encode a single open reading frame that is translated into one polyprotein. Viral proteinases process the viral polyprotein into precursor and mature proteins, resulting in the formation of four structural proteins and seven to eight mature non-structural proteins, depending on the virus. Viral proteins induce membrane rearrangements to form replication complexes, which are sites for RNA synthesis by the vRNA-dependent RNA polymerase, 3D. Newly synthesized RNAs undergo further rounds of translation/replication or become packaged into progeny virions. To redirect host resources toward virus replication, many picornaviruses rapidly shut down cap-dependent translation and disrupt nucleocytoplasmic trafficking to relocalize nuclear factors required for replication into the cytoplasm [1]. Modification of the cellular landscape is largely accomplished through the actions of non-structural proteins, particularly the viral proteinases 2A, 3C, and L (L is encoded by FMDV) [2,3].

Picornaviruses encounter multiple cellular processes that destabilize vRNA (Figure 1). Given their cytoplasmic replication cycle, picornaviruses avoid nuclear RNA surveillance mechanisms but instead are susceptible to mRNA decay pathways that function in the cytoplasm, such as adenylate uridylate-rich element (ARE)-mediated mRNA decay (AMD). Generally, mRNA decay is initiated by deadenylation of the targeted transcript, followed by 5′→3′ or 3′→5′ degradation of the body of the mRNA by exonucleases. To ensure that vRNAs are not targeted by mRNA decay machinery, picornaviruses disrupt these processes at multiple levels. The strategies that picornaviruses employ to disrupt this antiviral response have been researched more extensively than picornavirus involvement in mRNA decay pathways. This review highlights research focused on picornavirus interactions with pathways or processes associated with RNA decay, largely focusing on EVs, for which the most experimental evidence exists.

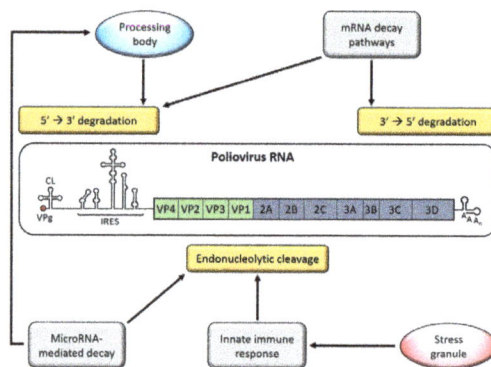

Figure 1. Poliovirus interactions with host RNA decay pathways. Picornavirus RNAs are exposed to degradation machinery from multiple RNA decay pathways and processes including 5′→3′ degradation in processing bodies (PBs), 5′→3′ or 3′→5′ degradation by mRNA decay pathways, and endonucleolytic cleavage as a result of activation of the innate immune response or microRNA-mediated decay. Poliovirus RNA serves as a model for the picornavirus genome. Genomic RNA is linked to a small virus-encoded protein, VPg, at the 5′ terminus. The 5′ non-coding region (NCR) contains a cloverleaf structure (CL) and an internal ribosome entry site (IRES). A single open reading frame codes both structural (VP1-4) and non-structural (2A-3D) proteins that are proteolytically processed into precursor and mature viral proteins. The 3′ end of the genome encodes a 3′ NCR and poly(A) tract.

2. Genome Stabilizing Features for Picornaviruses

Picornavirus RNAs have acquired several stabilizing features as a form of protection from cellular RNA decay machinery. At the 5′ end of the genome, the first barrier to decay machinery is a small, virus-encoded protein covalently bound to the 5′ terminal nucleotide of the vRNA, called VPg (Viral Protein, genome-linked) [4,5]. VPg is used by the vRNA polymerase, 3D, as a protein primer for RNA synthesis, resulting in vRNAs linked to VPg instead of a 7-methylguanosine (m7G) mRNA cap. Lacking a m7G cap, picornavirus RNAs are protected from cellular decapping enzymes like Dcp1 and Dcp2, whose activity initiates 5′→3′ RNA degradation. While cellular decapping enzymes are unable to hydrolyze the VPg-RNA bond, a cellular 5′-tyrosyl-DNA phosphodiesterase, TDP2, is capable of cleaving this bond and "unlinking" VPg from vRNA [6]. Cleavage of VPg may not result in destabilization of vRNAs since unlinked vRNA has been found to be associated with actively translating ribosomes and therefore protected from degradation, although the stability of unlinked vRNA has not yet been measured [7–9]. As an additional protection, the major 5′→3′ exonuclease, Xrn1, is degraded during poliovirus infection, preventing 5′→3′ digestion of unlinked vRNAs [10] (discussed further in Section 5).

For some picornaviruses, vRNA can inhibit endonucleolytic cleavage directly through association with ribonuclease L, (RNase L). RNase L activity is stimulated by the IFN response and contributes to the cellular defense against infection by degrading vRNA [11] (discussed further in Section 3). The antiviral activity of RNase L has been demonstrated for picornaviruses including EMCV and CVB4 [12,13]. However, poliovirus RNA is resistant to cleavage by RNase L through a structured RNA element located in the 3C proteinase coding sequence. This structured RNA binds the endoribonuclease domain of RNase L, which inhibits its activity [14]. This RNA element is conserved in group C EVs, which include poliovirus and several types of CVA, among others [15,16]. The element is not present in group A, B, or D EVs, of which CVB3 is a member and was found to be sensitive to RNase L. The protection of this RNA element from RNase L activity is not complete, however, as it was discovered that RNase L is still capable of cleaving the poliovirus genome at distinct locations [17].

Picornavirus RNA is stabilized through association with specific host proteins. Poly(rC)-binding protein 2 (PCBP2), an RNA-binding protein involved in mRNA stability and translation, binds 5′ stem-loop structures in poliovirus RNA to promote genomic RNA stability, viral translation, and RNA replication [18–20]. Mutation of one of these stem-loop structures to prevent PCBP2 binding results in diminished ability to form polysomes on poliovirus RNA, rendering the RNA susceptible to degradation [21,22]. PCBP2 also binds the 5′ NCR of CVB3, EV71 and HRV RNA to promote viral translation and RNA replication, but it has not yet been determined if this interaction affects vRNA stability [23–26]. Human antigen R (HuR) is a well-characterized mRNA stabilizing protein that was recently found to bind the EV71 5′ NCR and act as a positive regulator of translation [27]. While the effect of HuR on vRNA stability has yet to be determined, HuR may have a similar stabilizing effect as PCBP2, whereby promoting translation of vRNA protects it from degradation. HuR was previously shown to stabilize genomic RNAs of togaviruses, another family of positive-sense RNA viruses [28]. HuR was also identified as a poliovirus RNA-binding protein using thiouracil cross-linking mass spectrometry (TUX-MS), suggesting that it may have a similar effect on other vRNAs [29]. Host factors that promote picornavirus RNA stability have not been well studied, but it is likely that many of the proteins re-purposed for translation and replication serve a dual purpose in promoting vRNA stability.

3. Interferon (IFN)-Induced Viral RNA (vRNA) Degradation

The earliest defense against virus infection of cells involves activation of the innate immune response, which results in expression of genes that interfere with virus replication, prevent spread to neighboring cells, and trigger the adaptive immune response. Briefly described, detection of pathogen-associated molecular patterns (PAMPs) by cellular pattern recognition receptors (PRRs) initiates the innate immune response. During picornavirus infections, double-stranded RNAs (dsRNAs) that form during vRNA replication serve as the PAMP that is recognized by a PRR. PRR bound to viral dsRNA transduces the signal that a viral pathogen has been detected through multiple pathways, leading to the activation of transcription factors which promote the expression of IFN-β. IFN-β production ultimately results in transcription of hundreds of IFN-stimulated genes (ISGs) which collectively contribute to an antiviral state [30,31].

The IFN response activates multiple pathways to inhibit virus replication, including degradation of vRNA by RNase L. RNase L is normally expressed in mammalian cells, but remains inactive until infection is detected, resulting in ISG expression. Oligoadenylate synthetase (OAS) is an ISG that activates RNase L by generating 2′-5′ oligoadenylates (2-5A), the secondary messenger that induces dimerization and activation of RNase L [11,32]. Some picornaviruses have evolved mechanisms to directly inhibit RNase L activity. As noted above, RNase L can be directly inhibited by binding to a structured element within the RNA of group C EVs. Additionally, the TMEV L* protein has been shown to bind and inhibit RNase L [33]. The L* protein is an alternative, smaller form of L generated by leaky ribosome scanning [34]. In an uninfected cell, RNase L is inhibited by the cellular protein known as RNase L inhibitor/ATP-binding cassette, sub-family E member 1 (RLI/ABCE). EMCV infection induces RLI/ABCE expression, which contributes to RNase L inhibition [35]. It is not known whether other picornaviruses induce RLI/ABCE expression as well.

Several strategies are employed by picornaviruses to prevent endonucleolytic cleavage by RNase L prior to the nuclease becoming active. Melanoma differentiation-associated gene 5 (MDA5) is the PRR responsible for detecting the replicative form of picornavirus RNAs [36–42]. To avoid detection, MDA5 is cleaved or degraded during infection by poliovirus, CVB3, EV71, EMCV, or HRV1a. For poliovirus, CVB3, and EV71, the 2A proteinase appears to be responsible for cleaving MDA5; however, conflicting reports also indicate that cleavage may occur in a proteasome- or caspase-dependent manner [39,43,44]. Interestingly, MDA5 degradation is not common to all picornavirus infections, as MDA5 remains intact during HRV16 or echovirus type 1 infection [44]. Ligand-bound MDA5 assembles with the adaptor molecule mitochondrial antiviral signaling protein (MAVS) at the mitochondrial membrane to transfer the signal downstream. Accordingly, MAVS is also targeted for inhibition during viral infection. MAVS proteolysis has been observed during infection by poliovirus, CVB3, EV71, HRV1a, or HAV [43,45–47]. Several lines of evidence point toward MAVS cleavage by the 2A and/or 3C proteinases, as well as by caspases. In one study, expression of poliovirus, CVB3, or EV71 2A proteinase alone resulted in MDA5 and MAVS cleavage similar to what is observed during infection [43]. FMDV infection also results in reduction of MAVS, but not through its cleavage. Instead, the non-structural protein, 3A, and structural protein, VP3, have both been shown to down-regulate MAVS mRNA expression [48,49].

IFN-β expression is induced following recruitment of signaling molecules to MAVS which activates TANK-binding kinase 1 (TBK1), resulting in phosphorylation of the transcription factors IFN regulatory factor 3 and 7 (IRF-3 and -7). Phosphorylated IRF-3/7 proteins translocate to the nucleus to activate transcription of IFN-β. EV71 3C proteinase cleaves IRF-7, inhibiting its ability to transactivate IFN-β expression [50]. FMDV employs a different strategy to inhibit IFN-β expression. The FMDV Lb protein (generated by leaky ribosome scanning, similar to L*) deubiquitinates TBK1 and the signaling molecule TNF receptor-associated factor 3 (TRAF3), thereby inhibiting their activity [51]. In addition, the FMDV L protein causes a decrease in IRF-3/7 mRNA levels [52].

MAVS also activates proinflammatory cytokine expression by the nuclear factor-κB (NFκB) transcription factor. NFκB activation requires phosphorylation of the NFκB inhibitor-α (IκBα) by the IκB kinase complex, IKK (composed of IKKα, IKKβ, IKKγ). Phosphorylation of IκBα releases NFκB, allowing it to translocate to the nucleus and activate transcription. Various strategies are employed by picornaviruses to inhibit NFκB activation. Poliovirus, CVA16, CVB3 and EV71 2C proteins have been shown to inhibit phosphorylation and activation of IKK by recruiting protein phosphatase 1 (PP1) to IKKβ [53,54]. FMDV inhibits IKK through cleavage of IKKγ by 3C proteinase [55]. NFκB is also targeted directly through its p65/RelA subunit. Poliovirus, HRV14, echovirus type 1, and FMDV cleave p65/RelA during infection [56,57] and the EV71 2C protein has also been shown to inhibit NFκB by binding to p65/RelA [58].

Picornaviruses antagonize the innate immune response at many steps in the pathway, including steps not mentioned here (Figure 2). From preventing detection by vRNA sensors to inhibition and degradation of signal transduction molecules and transcription factors, picornaviruses employ a number of strategies to inhibit the antiviral response. Inhibition of this response promotes picornavirus replication and spread, in part by preventing the activation of RNase L, which poses a significant threat to vRNA stability.

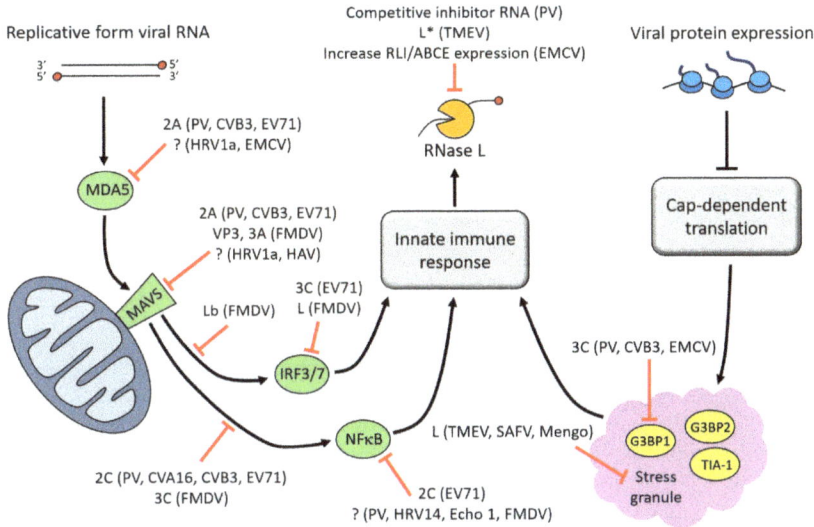

Figure 2. Inhibition of ribonuclease L (RNase L) activation by picornavirus proteins. RNase L is an effector molecule of the innate immune response. In its active form, RNase L endonucleolytically cleaves viral RNA (vRNA). Several picornaviruses have been shown to inhibit RNase L directly through binding of a viral inhibitor to RNase L or up-regulation of the cellular RNase L inhibitor, RLI/ABCE. Picornaviruses also indirectly prevent activation of RNase L through extensive disruption of pathways that contribute to the innate immune response. The signaling cascade that is initiated by MDA5 following detection of viral double-stranded RNA is disrupted at multiple steps during infection, which prevents expression of interferon (IFN)-induced genes and activation of RNase L. Additionally, the disassembly of stress granules (SGs) during infection inhibits SG-mediated enhancement of the innate immune response. Simplified pathways are depicted here, highlighting points in the pathway that are inhibited by specific picornaviruses and the viral protein responsible ("?" indicates that the viral protein responsible is unknown). PV, poliovirus; CVA16, coxsackievirus A16; CVB3, coxsackievirus B3; EV71, enterovirus 71; Echo 1, echovirus type 1; HRV1a, human rhinovirus 1a; EMCV, encephalomyocarditis virus; Mengo, mengovirus (a strain of EMCV); TMEV, Theiler's murine encephalomyelitis virus; SAFV, Saffold virus; HAV, hepatitis A virus; FMDV, foot-and-mouth disease virus.

4. Stress Granules (SGs)

Stress granules (SGs) are a type of cytoplasmic RNA granule that contain non-translating mRNAs and form in response to cellular stress and inhibition of translation [59–61]. Unlike processing bodies (PBs), which are RNA granules enriched for mRNA decay proteins, SGs contain many translation initiation factors and form around stalled translation initiation complexes [60,62–64]. SGs can assemble and disassemble continuously and may act as sites for mRNA storage and sorting between a repressed state, active translation on polysomes, or degradation in PBs [65–67]. Several mRNA decay factors associate with SGs, including the Xrn1 exonuclease [63], PMR1 endonuclease [68], and the mRNA decay proteins tristetraprolin (TTP), butyrate response factor 1 (BRF-1), and K homology-type splicing-regulatory protein (KSRP, also known as FBP2) [69,70]; however, SGs likely do not contribute to mRNA decay directly, but instead they promote degradation through their interaction with PBs. SGs and PBs can physically associate, share several protein components, and have been proposed to exchange "cargo," thereby targeting translationally repressed mRNAs for degradation [63,71]. Conversely, SGs are also thought to stabilize non-translating mRNAs by temporarily sequestering them away from components of the decay machinery [72,73]. In the context of viral infection, the major contribution to the destabilization of vRNA by SGs may be through enhancement of the innate immune response.

Several picornaviruses have been shown to transiently induce the formation of SGs early in infection, which disperse at roughly middle to late times during infection. The transient induction of SGs has been visualized by immunofluorescence for the EVs poliovirus [60,74,75], CVB3 [76], and EV71 [77], and cardioviruses EMCV [78] and TMEV [79]. During poliovirus infection, some virus-induced SGs are compositionally unique from stress-induced SGs, containing the splicing factor and viral IRES transacting factor (ITAF), SRSF3 (SRp20) [80]. A driving force behind picornavirus-induced SG formation is the shut-down of host cap-dependent translation. This leads to the accumulation of stalled translation initiation complexes which induce the aggregation of SGs. For that reason, expression of poliovirus, CVB3, or EV71 2A proteinases alone, which cleave the cap-binding complex component eIF4G, can induce SG formation [76,77,81].

Assembly of SGs is mediated by the SG-nucleating proteins Ras-Gap SH3 domain-binding protein 1 and 2 (G3BP1 and G3BP2) and T-cell-restricted intracellular antigen 1 (TIA-1), among others [82–84]. For poliovirus, CVB3, and EMCV, the disassembly of SGs occurs as a result of G3BP1 cleavage by 3C proteinase [74,76,78]. Expression of uncleavable G3BP1 prevents the disassembly of SGs during infection, highlighting the importance of 3C cleavage in disrupting virus-induced SGs [76,78]. TIA-1 and G3BP2 remain intact during infection, although that does not exclude cleavage of other SG-nucleating proteins from contributing to the disruption of SGs. Interestingly, G3BP1 cleavage does not contribute to the disassembly of SGs formed following infection by TMEV, which is in contrast to EMCV. Instead, TMEV-induced SGs are inhibited by the L protein through an unknown mechanism. The SAFV and mengovirus L proteins are also capable of inhibiting SGs when expressed in place of TMEV L [79]. Other viral proteins may also contribute to the inhibition of SGs. Expression of the poliovirus structural protein-coding region P1, 2A proteinase, or 3A alone modestly inhibited SGs induced by oxidative stress, although it is not understood whether these proteins play a part in inhibition of virus-induced SGs [81].

Picornavirus-induced SGs have previously been implicated in activation of the IFN response, but until recently it was not known how direct activation occurs. SGs induced by influenza A virus infection, a (−) ssRNA virus from the *Orthomyxoviridae* family, contain vRNA and several antiviral proteins including the vRNA sensors RIG-I and MDA5, along with OAS and RNase L. Formation of these antiviral SGs parallel, and potentially activate, the IFN response and concentrate vRNA in proximity to RNase L [85]. Similarly, transient SGs formed during EMCV infection were also associated with an antiviral effect. Expression of uncleavable G3BP1 prior to infection, which prevents the disassembly of virus-induced SGs, resulted in significantly increased levels of IFN-β and other cytokines, as well as decreased virus replication [78]. SGs induced by the overexpression of G3BP1 inhibit the replication of CVB3, CVB5, and EV70 and contain proteins involved in the innate immune response, such as OAS2, RNase L, and double-stranded RNA-dependent protein kinase (PKR) [86]. These data indicate that a link exists between SGs, G3BP1, and the innate immune response.

Recent investigations have revealed that G3BP1 directly stimulates the antiviral response through recruitment of PKR to SGs. It was shown that G3BP1 directly binds PKR in SGs during mengovirus infection, and in complex with another SG nucleating protein, Caprin1, activates PKR [87]. Active PKR then moves into the cytoplasm, where it can mediate the innate immune response through both its kinase activity and role as an adaptor protein. PKR induces IFN expression through activation of the transcription factor NFκB. PKR activation of NFκB has been shown to occur through indirect phosphorylation of the NFκB inhibitor IκB, which results in IκB degradation and the translocation of NFκB to the nucleus [88,89]. Interestingly, while PKR is required for IFN-α/β induction during EMCV and TMEV infection, it does not appear to do so through transcriptional activation, but instead through regulation of IFN mRNA integrity. EMCV infection of PKR$^{-/-}$ mouse cells resulted in very little IFN-β protein production despite normal IFN-β mRNA levels. Even though IFN-β mRNA levels appeared normal, the mRNA lacked a poly(A) tract, which resulted in its decreased translation [90].

Given a similar localization of PKR to EV-induced SGs, it would be reasonable to assume that PKR is activated in SGs formed by other picornaviruses, thereby enhancing the IFN response to infection. Even though RNase L was identified as a component of SGs formed following infection by CVB3, CVB5, or EV70, it is unlikely that this localization adds to the degradation of vRNA [87]. The proximity of RNase L to influenza A virus RNA in SGs was suggested as a possible mechanism for inhibition of virus replication; however, picornavirus RNA has been detected in SGs, which has been tested for poliovirus, CVB3, and TMEV [75,76,79]. Therefore, any contribution of virus-induced SGs to vRNA decay can most likely be attributed to the induction of IFNs, which results in activation of RNase L and degradation of vRNA. The disassembly of SGs during picornavirus infection plays an important part of maintaining vRNA stability through suppression of IFN-induced RNA decay (Figure 2).

5. Processing Bodies (PBs)

PBs are constitutively expressed cytoplasmic RNA granules which are composed of non-translating mRNAs and many proteins, including those involved in mRNA decay such as decapping components (Dcp1, Dcp2, Lsm1-7), $5' \rightarrow 3'$ exonuclease (Xrn1), and deadenylation factors (Pan2, Pan3, Ccr4, Caf1, poly(A) RNase (PARN)) [63,91–94]. Due to the concentration of mRNA decay machinery in PBs, these granules have been proposed to be involved in $5' \rightarrow 3'$ mRNA decay [93,95]. However, nonsense-mediated decay (NMD) and AMD have both been shown to occur in the absence of PB formation despite localization of NMD or AMD proteins in PBs [70,96]. This suggests that PBs are not required for all types of mRNA decay, but may form as a consequence of mRNA degradation or silencing and serve to enhance the process [97]. Not all mRNAs that enter PBs are degraded. mRNAs targeted for miRNA-mediated translational repression can localize to PBs [98–100]. Upon relief of repression, these mRNAs may leave the PB and re-enter active translation [101,102]. Given their complex role in mRNA decay and storage, it is not surprising that picornaviruses have evolved ways of disrupting PBs to prevent vRNAs from aggregating in these granules.

The effect of picornavirus infection on PBs has been studied for two EVs, poliovirus and CVB3. Following infection by poliovirus or CVB3, PB foci were almost completely absent by mid-infection. The PB proteins Dcp1a and Xrn1 are degraded during poliovirus infection. Loss of Dcp1a likely occurs as a result of poliovirus 3C cleavage, while Xrn1 does not appear to be a target of either viral proteinase and is instead degraded through a proteasome-dependent pathway [10]. Degradation of Xrn1 may have an additional benefit in protecting non-VPg-linked vRNAs from digestion by this major $5' \rightarrow 3'$ exonuclease. While cleavage and degradation of Dcp1a and Xrn1 likely contribute to PB disruption, the near-complete dispersal of PBs suggests that these granules are targeted through multiple mechanisms.

Deadenylation of mRNA is a necessary first step for PB formation. Inhibition of active deadenylation through the expression of dominant negative Caf1 results in almost complete disruption of PBs. Additionally, components of the deadenylase complexes Pan2–Pan3 and Ccr4–Caf1 localize to PBs, and siRNA-mediated knockdown of either Caf1, Ccr4, or Pan3 disrupt PB formation [94,103]. Therefore, inhibition of deadenylation presents an additional opportunity for virus-mediated disruption of PBs. Upon examination of deadenylase components following poliovirus infection, it was discovered that Pan3 is degraded, possibly as result of 3C cleavage, while PARN, Pan2, Ccr4, and Caf1 remain intact [10]. Knockdown of Pan3 has been shown to block PB formation but does not impair deadenylation of mRNAs [94]. Thus, poliovirus-mediated degradation of Pan3 may contribute to the disruption of PBs by preventing the association or nucleation of PB components, not through blocking deadenylation.

Several poliovirus proteins promote the dispersal of PBs when expressed individually, although the mechanisms for disruption remain unclear. Expression of either the 2A or 3C proteinases alone significantly reduced the number of PBs formed per cell, with 2A having a more pronounced effect than 3C. Additionally, expression of either 3CD (a precursor of 3C that also possesses proteinase activity) or the vRNA polymerase, 3D, induced a modest reduction in PBs, although their molecular targets are unknown. The dispersal of PBs by 2A or 3C appears to occur through different mechanisms. Expression of 2A could neither prevent the formation of stress-induced PBs nor disrupt PBs containing exogenously expressed Dcp1a, while 3C could do both [81]. These data emphasize the importance of the disassembly and prevention of formation of PBs during poliovirus and CVB3 infection, a process that likely extends to other picornaviruses. The disruption of PBs highlights the significance of this process in the virus life cycle (Figure 3).

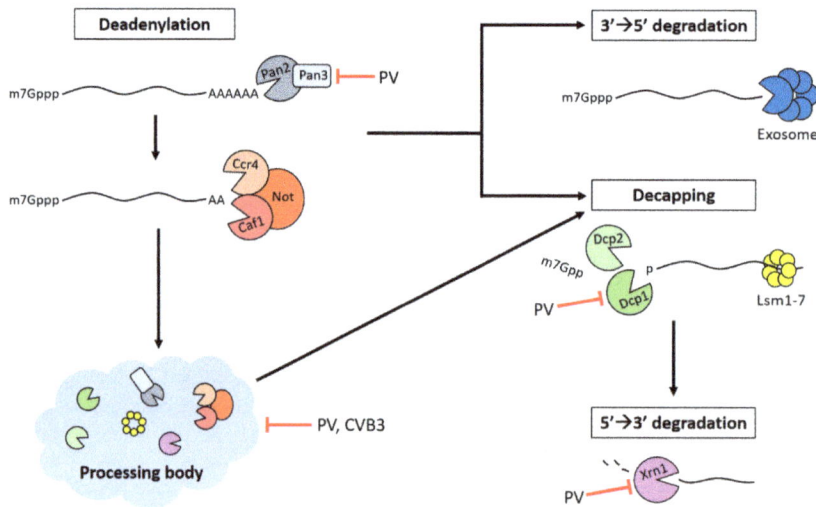

Figure 3. Inhibition of PBs and mRNA decay proteins during picornavirus infection. Deadenylation, decapping, and $5' \rightarrow 3'$ RNA degradation machinery are targeted during poliovirus (PV) and coxsackievirus B3 (CVB3) infection. The localization of non-translating RNAs and mRNA decay proteins in PBs indicate that these RNA granules may be involved in $5' \rightarrow 3'$ mRNA decay. PBs are dispersed during poliovirus and CVB3 infection, thereby disrupting their possible contribution to vRNA degradation. During poliovirus infection, Pan3 and Dcp1a are cleaved by 3C proteinase and Xrn1 is degraded by a proteasome-dependent mechanism.

6. Adenylate Uridylate-Rich Element (ARE)-Mediated mRNA Decay (AMD)

Regulation of mRNA stability and turnover is a critical component of post-transcriptional gene expression. Both *cis*- and *trans*-acting elements participate in regulating the stability of mRNA. Many mRNAs encode AREs, which are sequences often found within the 3′ UTR that regulate mRNA stability through their association with the approximately 20 identified ARE-binding proteins (AUBPs). Stabilization or degradation of a transcript depends on which AUBPs are bound. Several AUBPs have been well characterized for promoting AMD of target transcripts: ARE/poly(U)-binding/degradation factor 1 (AUF1), KSRP, TTP, and BRF-1/2. Other AUBPs are better known for either stabilizing (HuR, HuD) or repressing the translation (TIA-1, TIAR) of target transcripts [104].

AMD occurs in the cell cytoplasm and is initiated by deadenylase digestion of the 3′ poly(A) tract, followed by degradation of the body of the mRNA using both $3' \rightarrow 5'$ and $5' \rightarrow 3'$ exonucleolytic pathways [105–108]. Several decay-promoting AUBPs have been shown to directly interact with components of the RNA degradation machinery, suggesting a mechanism for their initiation of mRNA decay [109,110]. It seems plausible that binding of certain AUBPs to vRNA could lead to recruitment of RNA degradation machinery and subsequent decay of vRNA. If so, then AMD could present an additional antiviral strategy utilized by the cell. To date, two AUBPs known for promoting AMD, AUF1 and KSRP, have been characterized for their roles in the picornavirus life cycle. Both proteins are shown to have a negative impact on virus replication, but apparently not through their expected functions in promoting RNA decay.

AUF1 (also known as hnRNP D) is one of the best-described AUBPs involved in AMD and has been shown to promote the decay of mRNAs encoding oncogenic, inflammatory, and cell cycle proteins, among others [111–114]. In addition to its role in promoting mRNA decay, AUF1 has also been shown to stabilize [115,116] and promote the translation of targeted transcripts [117]. AUF1 is expressed as four different isoforms generated through alternative pre-mRNA splicing, named p37, p40, p42, and p45 based on their apparent molecular weights [118]. All of the AUF1 protein isoforms are composed of the same two, non-identical RNA recognition motifs (RRMs) and a glutamine-rich domain, but display different affinities for ARE substrates and subcellular localization [119,120]. The two largest isoforms of AUF1, p42 and p45, localize primarily to the nucleus, while the smaller p37 and p40 isoforms are predominantly nuclear, but transit between the nucleus and cytoplasm [121].

A large-scale identification of AUF1 target transcripts using photoactivatable ribonucleoside-enhanced crosslinking and immunoprecipitation (PAR-CLIP) revealed that AUF1 isoforms bind over 3000 mRNAs as well as many non-coding RNAs (ncRNAs) [122]. Pairing PAR-CLIP with RNA-seq and polysome analysis following AUF1 overexpression implicated AUF1 in a range of regulatory events including stabilization and destabilization of both mRNA and ncRNA, promotion and repression of translation, and modulation of alternative splicing. In addition, AUF1 has recently been shown to have a role in the life cycle of DNA and RNA viruses including Epstein-Barr virus [123,124], HIV [125], hepatitis C virus [126], West Nile virus [127,128], and several picornaviruses [129–132].

A direct interaction between AUF1 and picornavirus RNA was originally discovered from an RNA affinity screen using the 5′ NCR of poliovirus RNA [133]. This finding was later expanded to include binding to HRV, CVB3, and EV71 RNA [129,132,133]. It was subsequently discovered that AUF1 knockout or knockdown in human or mouse cells resulted in increased replication of these viruses, suggesting that AUF1 may act as a host restriction factor to EV infection [129–132]. Prior to these discoveries, a study of HRV-16 infection of human airway epithelial cells reported the observation that cytoplasmic levels of AUF1 increased during infection [134]. Cytoplasmic relocalization of AUF1 was also observed following poliovirus, HRV-14, CVB3, EV71, or EMCV infection of human cells [129–132]. For two of the viruses studied, poliovirus and CVB3, AUF1 was shown to relocalize as a result of the disruption of nuclear-cytoplasmic trafficking by the viral 2A proteinase [129,131]. Given the cytoplasmic life cycle of these viruses, relocalization of AUF1 to the cytoplasm may contribute significantly to its negative impact on virus replication.

Most reports aimed at determining the mechanism by which AUF1 acts as a restriction factor point toward a negative regulation of viral IRES-driven translation. The interaction of AUF1 with the 5′ NCR of vRNA has been demonstrated for poliovirus, HRV, and EV71, and this interaction was narrowed to sites within the poliovirus and EV71 IRES [129,132]. Using in vitro translation or bicistronic reporter assays, AUF1 was shown to negatively regulate both poliovirus and EV71 translation, likely through a direct interaction with the viral IRES [129,132]. For EV71, it was shown that the negative effect of AUF1 on viral translation may be a result of competitive binding to the viral IRES with a known ITAF and AUBP, hnRNP A1 [132,135,136]. hnRNP A1 also acts as an ITAF for HRV-2, but opposing regulation by hnRNP A1 and AUF1 has not yet been investigated for this virus [137]. An additional study has suggested that AUF1 may play a role in destabilizing CVB3 RNA. It was shown that AUF1 can bind directly to the CVB3 3′ NCR and that knockdown of AUF1 leads to increased stability of a CVB3 3′ NCR reporter construct [131]. Whether AUF1 promotes the decay of genomic CVB3 RNA has not yet been determined.

Like many host restriction factors, AUF1 is cleaved during infection. For poliovirus, HRV, and CVB3, AUF1 is cleaved by the 3C (and precursor 3CD) viral proteinase [131,133]. Cleavage of AUF1 by poliovirus 3CD proteinase, at a site within the *N*-terminal dimerization domain, reduces its ability to bind the 5′ NCR, which suggests that this process may act as a virus defense against AUF1 [129]. Interestingly, during EMCV infection of mouse cells, AUF1 did not have a negative impact on virus replication and remained uncleaved throughout infection [130]. Overall, these findings support the idea that AUF1 cleavage by viral proteinases ameliorates its negative effect on virus replication. However, many of the experiments demonstrating the negative effect of AUF1 on EV replication were performed using AUF1 knockout or knockdown cell lines. While a direct association with vRNA has been shown for several EVs, there may be additional indirect contributions to the negative effect of AUF1 resulting from the dysregulation of AUF1 target mRNA and ncRNA.

Similar to AUF1, KSRP is another AUBP that has been best characterized for its role in promoting AMD of target transcripts. Additional roles in transcription, alternative splicing, and miRNA maturation have also been described for KSRP [109,138–140]. Like AUF1, KSRP was identified in an RNA affinity screen for its association with vRNA; in this case, as a novel EV71 5′ NCR-binding protein. KSRP was shown to relocalize from the nucleus to the cytoplasm during infection and to associate with the EV71 5′ NCR, with binding occurring at multiple sites within the viral IRES. Using protein pulse-labeling and bicistronic reporter assays, KSRP was shown to be a negative ITAF for EV71 [141,142]. Like AUF1, KSRP is cleaved during infection. However, KSRP is not a substrate for the viral 2A or 3C proteinases but is instead cleaved and degraded through the activity of caspases and the proteasome and autophagy pathways [142]. KSRP has not yet been shown to act as a negative regulator of other picornavirus infections, but it was identified as a poliovirus RNA-binding protein using TUX-MS and thus, may have a similar effect on other EVs [29].

Other AUBPs have been shown to associate with picornavirus RNAs and participate in the picornavirus replication cycle; however, these proteins are not typically linked to AMD. Instead, these AUBPs are often associated with stabilization or translation of mRNA (discussed in Section 2). hnRNP A1, the AUBP that was shown to compete with AUF1 for binding to the EV71 IRES, is a multifunctional protein involved in transcription, alternative splicing, mRNA localization, translation, and stability [143]. hnRNP A1 has been reported to destabilize mRNAs bearing AREs [144] or an ARE-like motif (a motif which was identified in ~7% of mRNAs) [145]. However, instead of destabilizing EV71 RNA, hnRNP A1 is re-purposed by the virus as a positive regulator of translation [135]. Among its many functions, hnRNP A1 has also been shown to act as an ITAF for cellular IRESs and it is this function that is utilized to promote virus replication [135,146,147].

7. MicroRNA-Mediated Decay

MicroRNAs (miRNAs) are small, regulatory RNAs produced in eukaryotic cells that bind to complementary sites in mRNA and act to translationally repress or signal the degradation of target transcripts [148]. Biogenesis of miRNA begins in the nucleus, where precursor miRNAs (pre-miRNAs) are cleaved from hairpin structures within primary miRNA transcripts by the RNase III nuclease, Drosha [149]. Following export to the cytoplasm, pre-miRNAs are further processed to mature miRNAs by Dicer, another RNase III nuclease [150]. Mature miRNAs are bound by a member of the Argonaute protein family within the RNA-induced silencing complex (RISC), which together act as the effector complex that targets complementary mRNAs for RNA interference [151].

miRNA binding can lead directly to mRNA degradation, or indirectly through degradation of repressed mRNAs in PBs. Direct degradation of a miRNA target in animal cells is most often initiated through recruitment of deadenylases by the RISC, and on rare occasions, by endonucleolytic cleavage of the mRNA by Argonaute 2 (Ago2), the only catalytically active member of the Argonaute family [148,152]. Until a few years ago, it was assumed that antiviral potential of the miRNA pathway is not utilized during picornavirus infections, since the cytoplasmic replication cycle of these viruses does not encounter miRNA biogenesis pathways in the nucleus. However, regions of the viral

Viruses **2016**, *8*, 335

genome, such as the IRES, contain hairpin structures that resemble structured miRNA transcripts. These structured regions in the vRNA provide an opportunity for the cytoplasmic miRNA machinery to generate miRNA-like small RNAs.

Recent studies using deep sequencing techniques have revealed that viral small RNAs (vsRNAs) are produced from HAV, EMCV and EV71 RNA in a Dicer-dependent manner [153–155]. While the roles of the HAV and EMCV vsRNAs during infection have not been elucidated, the study of EV71 vsRNAs has revealed interesting new ways in which the virus re-purposes yet another potentially antiviral cellular pathway. One of the EV71 vsRNAs, vsRNA-1, is derived from the second stem loop (SL-II) of the 5′ NCR and negatively regulates viral translation and replication [155]. Instead of inhibiting virus replication via canonical miRNA mechanisms, it was discovered that vsRNA-1 may regulate EV71 IRES-driven translation by promoting the binding of both positive and negative ITAFs to SL-II of the 5′ NCR. Surprisingly, vsRNA-1 recruits Ago2 to the vRNA, but instead of acting as a negative regulator through translational repression or cleavage of vRNA, Ago2 acts as a positive regulator of translation [27]. These data reveal the possibility for a new and interesting regulator in the picornavirus life cycle. Whether these small, virus-derived RNAs have a negative impact on vRNA stability similar to miRNA has yet to be determined.

8. Concluding Remarks

Picornavirus disruption of cellular RNA decay machinery generally involves a broad approach of cleaving, degrading, inhibiting, disassembling, or re-purposing components of these processes. However, several pathways that may contribute to vRNA instability have received relatively little attention to date, and the focus has been on only a few members of the *Picornaviridae* family. For instance, 5′→3′ RNA degradation is inhibited during infection through cleavage of decapping enzymes and Xrn1, but it is unclear whether exosomes, which participate in 3′→5′ RNA degradation, remain intact. There has also been little investigation into the cellular proteins that bind vRNA and which promote vRNA stability, or conversely, promote degradation. Insight into these issues can be provided by approaches like TUX-MS, which may reveal novel RNA-binding proteins that contribute to vRNA stability. In addition, studying the picornavirus "cleavome," or host proteins that are cleaved by viral proteinases during infection, may reveal new proteins whose functions are disrupted during infection. These cleaved proteins could include negative regulators of vRNA stability. Studies like these will reveal not only new viral mechanisms to preserve the stability of their encoded RNAs, but also host defense mechanisms that target vRNA for degradation.

Acknowledgments: We are indebted to Autumn Holmes, Dylan Flather, and Eric Baggs for their editorial contributions to this work. Work from the authors' lab described in this review was supported by U.S. Public Health Service grants AI022693, AI026765, and AI110782 from the National Institutes of Health (to B.L.S.). W.U. is a predoctoral trainee supported by U.S. Public Health Service training grant AI007319 from the National Institutes of Health.

Author Contributions: W.U. researched and wrote the manuscript. B.L.S. provided critical input into the design, writing, and editorial revisions of the manuscript.

Conflicts of Interest: The authors declare no conflict of interest.

References

1. Flather, D.; Semler, B.L. Picornaviruses and nuclear functions: Targeting a cellular compartment distinct from the replication site of a positive-strand RNA virus. *Front. Microbiol.* **2015**, *6*, 594. [CrossRef] [PubMed]
2. Lin, J.Y.; Chen, T.C.; Weng, K.F.; Chang, S.C.; Chen, L.L.; Shih, S.R. Viral and host proteins involved in picornavirus life cycle. *J. Biomed. Sci.* **2009**, *16*, 103. [CrossRef] [PubMed]
3. Racaniello, V.R. *Picornaviridae*: The viruses and their replication. In *Fields Virology*, 6th ed.; Knipe, D.M., Howley, P.M., Eds.; Lippincott Williams & Wilkins: Philadelphia, PA, USA, 2013; pp. 453–489.
4. Lee, Y.F.; Nomoto, A.; Detjen, B.M.; Wimmer, E. A protein covalently linked to poliovirus genome RNA. *Proc. Natl. Acad. Sci. USA* **1977**, *74*, 59–63. [CrossRef] [PubMed]

5. Flanegan, J.B.; Petterson, R.F.; Ambros, V.; Hewlett, N.J.; Baltimore, D. Covalent linkage of a protein to a defined nucleotide sequence at the 5'-terminus of virion and replicative intermediate RNAs of poliovirus. *Proc. Natl. Acad. Sci. USA* **1977**, *74*, 961–965. [CrossRef] [PubMed]

6. Virgen-Slane, R.; Rozovics, J.M.; Fitzgerald, K.D.; Ngo, T.; Chou, W.; van der Heden van Noort, G.J.; Filippov, D.V.; Gershon, P.D.; Semler, B.L. An RNA virus hijacks an incognito function of a DNA repair enzyme. *Proc. Natl. Acad. Sci. USA* **2012**, *109*, 14634–14639. [CrossRef] [PubMed]

7. Nomoto, A.; Lee, Y.F.; Wimmer, E. The 5' end of poliovirus mRNA is not capped with m7G(5')ppp(5')Np. *Proc. Natl. Acad. Sci. USA* **1976**, *73*, 375–380. [CrossRef] [PubMed]

8. Fernandez-Munoz, R.; Darnell, J.E. Structural difference between the 5' termini of viral and cellular mRNA in poliovirus-infected cells: Possible basis for the inhibition of host protein synthesis. *J. Virol.* **1976**, *18*, 719–726. [PubMed]

9. Hewlett, M.J.; Rose, J.K.; Baltimore, D. 5'-terminal structure of poliovirus polyribosomal RNA is pUp. *Proc. Natl. Acad. Sci. USA* **1976**, *73*, 327–330. [CrossRef] [PubMed]

10. Dougherty, J.D.; White, J.P.; Lloyd, R.E. Poliovirus-mediated disruption of cytoplasmic processing bodies. *J. Virol.* **2011**, *85*, 64–75. [CrossRef] [PubMed]

11. Drappier, M.; Michiels, T. Inhibition of the OAS/RNase L pathway by viruses. *Curr. Opin. Virol.* **2015**, *15*, 19–26. [CrossRef] [PubMed]

12. Flodström-Tullberg, M.; Hultcrantz, M.; Stotland, A.; Maday, A.; Tsai, D.; Fine, C.; Williams, B.; Silverman, R.; Sarvetnick, N. RNase L and double-stranded RNA-dependent protein kinase exert complementary roles in islet cell defense during coxsackievirus infection. *J. Immunol.* **2005**, *174*, 1171–1177. [CrossRef] [PubMed]

13. Coccia, E.M.; Romeo, G.; Nissim, A.; Marziali, G.; Albertini, R.; Affabris, E.; Battistini, A.; Fiorucci, G.; Orsatti, R.; Rossi, G.B.; et al. A full-length murine 2-5A synthetase cDNA transfected in NIH-3T3 cells impairs EMCV but not VSV replication. *Virology* **1990**, *179*, 228–233. [CrossRef]

14. Townsend, H.L.; Jha, B.K.; Han, J.-Q.; Maluf, N.K.; Silverman, R.H.; Barton, D.J. A viral RNA competitively inhibits the antiviral endoribonuclease domain of RNase L. *RNA* **2008**, *14*, 1026–1036. [CrossRef] [PubMed]

15. Han, J.-Q.; Townsend, H.L.; Jha, B.K.; Paranjape, J.M.; Silverman, R.H.; Barton, D.J. A phylogenetically conserved RNA structure in the poliovirus open reading frame inhibits the antiviral endoribonuclease RNase L. *J. Virol.* **2007**, *81*, 5561–5572. [CrossRef] [PubMed]

16. Townsend, H.L.; Jha, B.K.; Silverman, R.H.; Barton, D.J. A putative loop E motif and an H-H kissing loop interaction are conserved and functional features in a group C enterovirus RNA that inhibits ribonuclease L. *RNA Biol.* **2008**, *5*, 263–272. [CrossRef] [PubMed]

17. Cooper, D.A.; Jha, B.K.; Silverman, R.H.; Hesselberth, J.R.; Barton, D.J. Ribonuclease L and metal-ion–independent endoribonuclease cleavage sites in host and viral RNAs. *Nucleic Acids Res.* **2014**, *42*, 5202–5216. [CrossRef] [PubMed]

18. Blyn, L.B.; Swiderek, K.M.; Richards, O.; Stahl, D.C.; Semler, B.L.; Ehrenfeld, E. Poly(rC) binding protein 2 binds to stem-loop IV of the poliovirus RNA 5' noncoding region: Identification by automated liquid chromatography-tandem mass spectrometry. *Proc. Natl. Acad. Sci. USA* **1996**, *93*, 11115–11120. [CrossRef] [PubMed]

19. Parsley, T.B.; Towner, J.S.; Blyn, L.B.; Ehrenfeld, E.; Semler, B.L. Poly (rC) binding protein 2 forms a ternary complex with the 5'-terminal sequences of poliovirus RNA and the viral 3CD proteinase. *RNA* **1997**, *3*, 1124–1134. [PubMed]

20. Gamarnik, A.V.; Andino, R. Two functional complexes formed by KH domain containing proteins with the 5' noncoding region of poliovirus RNA. *RNA* **1997**, *3*, 882–892. [PubMed]

21. Murray, K.E.; Roberts, A.W.; Barton, D.J. Poly(rC) binding proteins mediate poliovirus mRNA stability. *RNA* **2001**, *7*, 1126–1141. [CrossRef] [PubMed]

22. Kempf, B.J.; Barton, D.J. Poly(rC) binding proteins and the 5' cloverleaf of uncapped poliovirus mRNA function during de novo assembly of polysomes. *J. Virol.* **2008**, *82*, 5835–5846. [CrossRef] [PubMed]

23. Zell, R.; Ihle, Y.; Seitz, S.; Gündel, U.; Wutzler, P.; Görlach, M. Poly(rC)-binding protein 2 interacts with the oligo(rC) tract of coxsackievirus B3. *Biochem. Biophys. Res. Commun.* **2008**, *366*, 917–921. [CrossRef] [PubMed]

24. Hunt, S.L.; Hsuan, J.J.; Totty, N.; Jackson, R.J. unr, a cellular cytoplasmic RNA-binding protein with five cold-shock domains, is required for internal initiation of translation of human rhinovirus RNA. *Genes Dev.* **1999**, *13*, 437–448. [CrossRef] [PubMed]

25. Lin, J.-Y.; Li, M.-L.; Huang, P.-N.; Chien, K.-Y.; Horng, J.-T.; Shih, S.-R. Heterogeneous nuclear ribonuclear protein K interacts with the enterovirus 71 5′ untranslated region and participates in virus replication. *J. Gen. Virol.* **2008**, *89*, 2540–2549. [CrossRef] [PubMed]

26. Walter, B.L.; Nguyen, J.H.; Ehrenfeld, E.; Semler, B.L. Differential utilization of poly(rC) binding protein 2 in translation directed by picornavirus IRES elements. *RNA* **1999**, *5*, 1570–1585. [CrossRef] [PubMed]

27. Lin, J.-Y.; Brewer, G.; Li, M.-L. HuR and Ago2 bind the internal ribosome entry site of enterovirus 71 and promote virus translation and replication. *PLoS ONE* **2015**, *10*, e0140291. [CrossRef] [PubMed]

28. Sokoloski, K.J.; Dickson, A.M.; Chaskey, E.L.; Garneau, N.L.; Wilusz, C.J.; Wilusz, J. Sindbis virus usurps the cellular HuR protein to stabilize its transcripts and promote productive infections in mammalian and mosquito cells. *Cell Host Microbe* **2010**, *8*, 196–207. [CrossRef] [PubMed]

29. Lenarcic, E.M.; Landry, D.M.; Greco, T.M.; Cristea, I.M.; Thompson, S.R. Thiouracil cross-linking mass spectrometry: A cell-based method to identify host factors involved in viral amplification. *J. Virol.* **2013**, *87*, 8697–8712. [CrossRef] [PubMed]

30. Feng, Q.; Langereis, M.A.; van Kuppeveld, F.J.M. Induction and suppression of innate antiviral responses by picornaviruses. *Cytokine Growth Factor Rev.* **2014**, *25*, 577–585. [CrossRef] [PubMed]

31. Lu, J.; Yi, L.; Ke, C.; Zhang, Y.; Liu, R.; Chen, J.; Kung, H.-f.; He, M.-L. The interaction between human enteroviruses and type I IFN signaling pathway. *Crit. Rev. Microbiol.* **2015**, *41*, 201–207. [CrossRef] [PubMed]

32. Bisbal, C.; Silverman, R.H. Diverse functions of RNase L and implications in pathology. *Biochimie* **2007**, *89*, 789–798. [CrossRef] [PubMed]

33. Sorgeloos, F.; Jha, B.K.; Silverman, R.H.; Michiels, T. Evasion of antiviral innate immunity by Theiler's virus L* protein through direct inhibition of RNase L. *PLoS Pathog.* **2013**, *9*, e1003474. [CrossRef] [PubMed]

34. Chen, H.H.; Kong, W.P.; Zhang, L.; Ward, P.L.; Roos, R.P. A picornaviral protein synthesized out of frame with the polyprotein plays a key role in a virus-induced immune-mediated demyelinating disease. *Nat. Med.* **1995**, *1*, 927–931. [CrossRef] [PubMed]

35. Martinand, C.; Salehzada, T.; Silhol, M.; Lebleu, B.; Bisbal, C. RNase L inhibitor (RLI) antisense constructions block partially the down regulation of the 2-5A/RNase L pathway in encephalomyocarditis-virus-(EMCV)-infected cells. *Eur. J. Biochem.* **1998**, *254*, 248–255. [CrossRef] [PubMed]

36. Gitlin, L.; Barchet, W.; Gilfillan, S.; Cella, M.; Beutler, B.; Flavell, R.A.; Diamond, M.S.; Colonna, M. Essential role of MDA-5 in type I IFN responses to polyriboinosinic:polyribocytidylic acid and encephalomyocarditis picornavirus. *Proc. Natl. Acad. Sci. USA* **2006**, *103*, 8459–8464. [CrossRef] [PubMed]

37. Jin, Y.-H.; Kim, S.J.; So, E.Y.; Meng, L.; Colonna, M.; Kim, B.S. Melanoma differentiation-associated gene 5 is critical for protection against Theiler's virus-induced demyelinating disease. *J. Virol.* **2012**, *86*, 1531–1543. [CrossRef] [PubMed]

38. Kato, H.; Takeuchi, O.; Sato, S.; Yoneyama, M.; Yamamoto, M.; Matsui, K.; Uematsu, S.; Jung, A.; Kawai, T.; Ishii, K.J.; et al. Differential roles of MDA5 and RIG-I helicases in the recognition of RNA viruses. *Nature* **2006**, *441*, 101–105. [CrossRef] [PubMed]

39. Kuo, R.-L.; Kao, L.-T.; Lin, S.-J.; Wang, R.Y.-L.; Shih, S.-R. MDA5 plays a crucial role in enterovirus 71 RNA-mediated IRF3 activation. *PLoS ONE* **2013**, *8*, e63431. [CrossRef] [PubMed]

40. Wang, J.P.; Cerny, A.; Asher, D.R.; Kurt-Jones, E.A.; Bronson, R.T.; Finberg, R.W. MDA5 and MAVS mediate type I interferon responses to coxsackie B virus. *J. Virol.* **2010**, *84*, 254–260. [CrossRef] [PubMed]

41. Feng, Q.; Hato, S.V.; Langereis, M.A.; Zoll, J.; Virgen-Slane, R.; Peisley, A.; Hur, S.; Semler, B.L.; van Rij, R.P.; van Kuppeveld, F.J. MDA5 detects the double-stranded RNA replicative form in picornavirus-infected cells. *Cell Rep.* **2012**, *2*, 1187–1196. [CrossRef] [PubMed]

42. Abe, Y.; Fujii, K.; Nagata, N.; Takeuchi, O.; Akira, S.; Oshiumi, H.; Matsumoto, M.; Seya, T.; Koike, S. The toll-like receptor 3-mediated antiviral response is important for protection against poliovirus infection in poliovirus receptor transgenic mice. *J. Virol.* **2012**, *86*, 185–194. [CrossRef] [PubMed]

43. Feng, Q.; Langereis, M.A.; Lork, M.; Nguyen, M.; Hato, S.V.; Lanke, K.; Emdad, L.; Bhoopathi, P.; Fisher, P.B.; Lloyd, R.E.; et al. Enterovirus 2Apro targets MDA5 and MAVS in infected cells. *J. Virol.* **2014**, *88*, 3369–3378. [CrossRef] [PubMed]

44. Barral, P.M.; Morrison, J.M.; Drahos, J.; Gupta, P.; Sarkar, D.; Fisher, P.B.; Racaniello, V.R. MDA-5 is cleaved in poliovirus-infected cells. *J. Virol.* **2007**, *81*, 3677–3684. [CrossRef] [PubMed]

45. Drahos, J.; Racaniello, V.R. Cleavage of IPS-1 in cells infected with human rhinovirus. *J. Virol.* **2009**, *83*, 11581–11587. [CrossRef] [PubMed]

46. Wang, B.; Xi, X.; Lei, X.; Zhang, X.; Cui, S.; Wang, J.; Jin, Q.; Zhao, Z. Enterovirus 71 protease 2A^{pro} targets MAVS to inhibit anti-viral type I interferon responses. *PLoS Pathog.* **2013**, *9*, e1003231. [CrossRef] [PubMed]

47. Yang, Y.; Liang, Y.; Qu, L.; Chen, Z.; Yi, M.; Li, K.; Lemon, S.M. Disruption of innate immunity due to mitochondrial targeting of a picornaviral protease precursor. *Proc. Natl. Acad. Sci. USA* **2007**, *104*, 7253–7258. [CrossRef] [PubMed]

48. Li, D.; Yang, W.; Yang, F.; Liu, H.; Zhu, Z.; Lian, K.; Lei, C.; Li, S.; Liu, X.; Zheng, H.; et al. The VP3 structural protein of foot-and-mouth disease virus inhibits the IFN-β signaling pathway. *FASEB J.* **2016**, *30*, 1757–1766. [CrossRef] [PubMed]

49. Li, D.; Lei, C.; Xu, Z.; Yang, F.; Liu, H.; Zhu, Z.; Li, S.; Liu, X.; Shu, H.; Zheng, H. Foot-and-mouth disease virus non-structural protein 3A inhibits the interferon-β signaling pathway. *Sci. Rep.* **2016**, *6*, 21888. [CrossRef] [PubMed]

50. Lei, X.; Xiao, X.; Xue, Q.; Jin, Q.; He, B.; Wang, J. Cleavage of interferon regulatory factor 7 by enterovirus 71 3C suppresses cellular responses. *J. Virol.* **2013**, *87*, 1690–1698. [CrossRef] [PubMed]

51. Wang, D.; Fang, L.; Li, P.; Sun, L.; Fan, J.; Zhang, Q.; Luo, R.; Liu, X.; Li, K.; Chen, H.; et al. The leader proteinase of foot-and-mouth disease virus negatively regulates the type I interferon pathway by acting as a viral deubiquitinase. *J. Virol.* **2011**, *85*, 3758–3766. [CrossRef] [PubMed]

52. Wang, D.; Fang, L.; Luo, R.; Ye, R.; Fang, Y.; Xie, L.; Chen, H.; Xiao, S. Foot-and-mouth disease virus leader proteinase inhibits dsRNA-induced type I interferon transcription by decreasing interferon regulatory factor 3/7 in protein levels. *Biochem. Biophys. Res. Commun.* **2010**, *399*, 72–78. [CrossRef] [PubMed]

53. Li, Q.; Zheng, Z.; Liu, Y.; Zhang, Z.; Liu, Q.; Meng, J.; Ke, X.; Hu, Q.; Wang, H. 2C proteins of enteroviruses suppress IKKβ phosphorylation by recruiting protein phosphatase 1. *J. Virol.* **2016**, *90*, 5141–5151. [CrossRef] [PubMed]

54. Zheng, Z.; Li, H.; Zhang, Z.; Meng, J.; Mao, D.; Bai, B.; Lu, B.; Mao, P.; Hu, Q.; Wang, H. Enterovirus 71 2C protein inhibits TNF-α–mediated activation of NF-κB by suppressing IκB kinase β phosphorylation. *J. Immunol.* **2011**, *187*, 2202–2212. [CrossRef] [PubMed]

55. Wang, D.; Fang, L.; Li, K.; Zhong, H.; Fan, J.; Ouyang, C.; Zhang, H.; Duan, E.; Luo, R.; Zhang, Z.; et al. Foot-and-mouth disease virus 3C protease cleaves NEMO to impair innate immune signaling. *J. Virol.* **2012**, *86*, 9311–9322. [CrossRef] [PubMed]

56. De Los Santos, T.; Diaz-San Segundo, F.; Grubman, M.J. Degradation of nuclear factor kappa B during foot-and-mouth disease virus infection. *J. Virol.* **2007**, *81*, 12803–12815. [CrossRef] [PubMed]

57. Neznanov, N.; Chumakov, K.M.; Neznanova, L.; Almasan, A.; Banerjee, A.K.; Gudkov, A.V. Proteolytic cleavage of the p65-RelA subunit of NF-κB during poliovirus infection. *J. Biol. Chem.* **2005**, *280*, 24153–24158. [CrossRef] [PubMed]

58. Du, H.; Yin, P.; Yang, X.; Zhang, L.; Jin, Q.; Zhu, G. Enterovirus 71 2C protein inhibits NF-κB activation by binding to RelA(p65). *Sci. Rep.* **2015**, *5*, 14302. [CrossRef] [PubMed]

59. Kedersha, N.L.; Gupta, M.; Li, W.; Miller, I.; Anderson, P. RNA-binding Proteins TIA-1 and TIAR link the phosphorylation of eIF-2α to the assembly of mammalian stress granules. *J. Cell Biol.* **1999**, *147*, 1431–1442. [CrossRef] [PubMed]

60. Mazroui, R.; Sukarieh, R.; Bordeleau, M.-E.; Kaufman, R.J.; Northcote, P.; Tanaka, J.; Gallouzi, I.; Pelletier, J. Inhibition of ribosome recruitment induces stress granule formation independently of eukaryotic initiation factor 2α phosphorylation. *Mol. Biol. Cell* **2006**, *17*, 4212–4219. [CrossRef] [PubMed]

61. McEwen, E.; Kedersha, N.; Song, B.; Scheuner, D.; Gilks, N.; Han, A.; Chen, J.-J.; Anderson, P.; Kaufman, R.J. Heme-regulated inhibitor kinase-mediated phosphorylation of eukaryotic translation initiation factor 2 inhibits translation, induces stress granule formation, and mediates survival upon arsenite exposure. *J. Biol. Chem.* **2005**, *280*, 16925–16933. [CrossRef] [PubMed]

62. Kedersha, N.; Anderson, P. Stress granules: Sites of mRNA triage that regulate mRNA stability and translatability. *Biochem. Soc.Trans.* **2002**, *30*, 963–969. [CrossRef] [PubMed]

63. Kedersha, N.; Stoecklin, G.; Ayodele, M.; Yacono, P.; Lykke-Andersen, J.; Fritzler, M.J.; Scheuner, D.; Kaufman, R.J.; Golan, D.E.; Anderson, P. Stress granules and processing bodies are dynamically linked sites of mRNP remodeling. *J. Cell Biol.* **2005**, *169*, 871–884. [CrossRef] [PubMed]

64. Kimball, S.R.; Horetsky, R.L.; Ron, D.; Jefferson, L.S.; Harding, H.P. Mammalian stress granules represent sites of accumulation of stalled translation initiation complexes. *Am. J. Physiol. Cell Physiol.* **2003**, *284*, C273–C284. [CrossRef] [PubMed]

65. Hilliker, A.; Gao, Z.; Jankowsky, E.; Parker, R. The DEAD-box protein Ded1 modulates translation by the formation and resolution of an eIF4F-mRNA complex. *Mol. Cell* **2011**, *43*, 962–972. [CrossRef] [PubMed]

66. Kedersha, N.; Cho, M.R.; Li, W.; Yacono, P.W.; Chen, S.; Gilks, N.; Golan, D.E.; Anderson, P. Dynamic shuttling of TIA-1 accompanies the recruitment of mRNA to mammalian stress granules. *J. Cell Biol.* **2000**, *151*, 1257–1268. [CrossRef] [PubMed]

67. Mollet, S.; Cougot, N.; Wilczynska, A.; Dautry, F.; Kress, M.; Bertrand, E.; Weil, D. Translationally repressed mRNA transiently cycles through stress granules during stress. *Mol. Biol. Cell* **2008**, *19*, 4469–4479. [CrossRef] [PubMed]

68. Yang, F.; Peng, Y.; Murray, E.L.; Otsuka, Y.; Kedersha, N.; Schoenberg, D.R. Polysome-bound endonuclease PMR1 is targeted to stress granules via stress-specific binding to TIA-1. *Mol. Cell. Biol.* **2006**, *26*, 8803–8813. [CrossRef] [PubMed]

69. Rothé, F.; Gueydan, C.; Bellefroid, E.; Huez, G.; Kruys, V. Identification of FUSE-binding proteins as interacting partners of TIA proteins. *Biochem. Biophys. Res. Commun.* **2006**, *343*, 57–68. [CrossRef] [PubMed]

70. Stoecklin, G.; Stubbs, T.; Kedersha, N.; Wax, S.; Rigby, W.F.; Blackwell, T.K.; Anderson, P. MK2-induced tristetraprolin:14-3-3 complexes prevent stress granule association and ARE-mRNA decay. *EMBO J.* **2004**, *23*, 1313–1324. [CrossRef] [PubMed]

71. Wilczynska, A.; Aigueperse, C.; Kress, M.; Dautry, F.; Weil, D. The translational regulator CPEB1 provides a link between dcp1 bodies and stress granules. *J. Cell Sci.* **2005**, *118*, 981–992. [CrossRef] [PubMed]

72. Gareau, C.; Fournier, M.-J.; Filion, C.; Coudert, L.; Martel, D.; Labelle, Y.; Mazroui, R. p21WAF1/CIP1 upregulation through the stress granule-associated protein CUGBP1 confers resistance to bortezomib-mediated apoptosis. *PLoS ONE* **2011**, *6*, e20254. [CrossRef] [PubMed]

73. Mazroui, R.; Di Marco, S.; Kaufman, R.J.; Gallouzi, I.-E. Inhibition of the ubiquitin-proteasome system induces stress granule formation. *Mol. Biol. Cell* **2007**, *18*, 2603–2618. [CrossRef] [PubMed]

74. White, J.P.; Cardenas, A.M.; Marissen, W.E.; Lloyd, R.E. Inhibition of cytoplasmic mRNA stress granule formation by a viral proteinase. *Cell Host Microbe* **2007**, *2*, 295–305. [CrossRef] [PubMed]

75. Piotrowska, J.; Hansen, S.J.; Park, N.; Jamka, K.; Sarnow, P.; Gustin, K.E. Stable formation of compositionally unique stress granules in virus-infected cells. *J. Virol.* **2010**, *84*, 3654–3665. [CrossRef] [PubMed]

76. Fung, G.; Ng, C.S.; Zhang, J.; Shi, J.; Wong, J.; Piesik, P.; Han, L.; Chu, F.; Jagdeo, J.; Jan, E.; et al. Production of a dominant-negative fragment due to G3BP1 cleavage contributes to the disruption of mitochondria-associated protective stress granules during CVB3 infection. *PLoS ONE* **2013**, *8*, e79546. [CrossRef] [PubMed]

77. Wu, S.; Wang, Y.; Lin, L.; Si, X.; Wang, T.; Zhong, X.; Tong, L.; Luan, Y.; Chen, Y.; Li, X.; et al. Protease 2A induces stress granule formation during coxsackievirus B3 and enterovirus 71 infections. *Virol. J.* **2014**, *11*, 1–10. [CrossRef] [PubMed]

78. Ng, C.S.; Jogi, M.; Yoo, J.-S.; Onomoto, K.; Koike, S.; Iwasaki, T.; Yoneyama, M.; Kato, H.; Fujita, T. Encephalomyocarditis virus disrupts stress granules, the critical platform for triggering antiviral innate immune responses. *J. Virol.* **2013**, *87*, 9511–9522. [CrossRef] [PubMed]

79. Borghese, F.; Michiels, T. The leader protein of cardioviruses inhibits stress granule assembly. *J. Virol.* **2011**, *85*, 9614–9622. [CrossRef] [PubMed]

80. Fitzgerald, K.D.; Semler, B.L. Poliovirus infection induces the co-localization of cellular protein SRp20 with TIA-1, a cytoplasmic stress granule protein. *Virus Res.* **2013**, *176*, 223–231. [CrossRef] [PubMed]

81. Dougherty, J.; Tsai, W.-C.; Lloyd, R. Multiple poliovirus proteins repress cytoplasmic RNA granules. *Viruses* **2015**, *7*, 2922. [CrossRef] [PubMed]

82. Gilks, N.; Kedersha, N.; Ayodele, M.; Shen, L.; Stoecklin, G.; Dember, L.M.; Anderson, P. Stress granule assembly is mediated by prion-like aggregation of TIA-1. *Mol. Biol. Cell* **2004**, *15*, 5383–5398. [CrossRef] [PubMed]

83. Matsuki, H.; Takahashi, M.; Higuchi, M.; Makokha, G.N.; Oie, M.; Fujii, M. Both G3BP1 and G3BP2 contribute to stress granule formation. *Genes Cells* **2013**, *18*, 135–146. [CrossRef] [PubMed]

84. Tourrière, H.; Chebli, K.; Zekri, L.; Courselaud, B.; Blanchard, J.M.; Bertrand, E.; Tazi, J. The RasGAP-associated endoribonuclease G3BP assembles stress granules. *J. Cell Biol.* **2003**, *160*, 823–831. [CrossRef] [PubMed]

85. Onomoto, K.; Jogi, M.; Yoo, J.-S.; Narita, R.; Morimoto, S.; Takemura, A.; Sambhara, S.; Kawaguchi, A.; Osari, S.; Nagata, K.; et al. Critical role of an antiviral stress granule containing RIG-I and PKR in viral detection and innate immunity. *PLoS ONE* **2012**, *7*, e43031. [CrossRef]

86. Reineke, L.C.; Lloyd, R.E. The stress granule protein G3BP1 recruits protein kinase R to promote multiple innate immune antiviral responses. *J. Virol.* **2015**, *89*, 2575–2589. [CrossRef] [PubMed]
87. Reineke, L.C.; Kedersha, N.; Langereis, M.A.; van Kuppeveld, F.J.; Lloyd, R.E. Stress granules regulate double-stranded RNA-dependent protein kinase activation through a complex containing G3BP1 and Caprin1. *mBio* **2015**, *6*, e02486-14. [CrossRef] [PubMed]
88. Gil, J.; García, M.A.; Gomez-Puertas, P.; Guerra, S.; Rullas, J.; Nakano, H.; Alcamí, J.; Esteban, M. TRAF family proteins link PKR with NF-κB activation. *Mol. Cell. Biol.* **2004**, *24*, 4502–4512. [CrossRef] [PubMed]
89. Pindel, A.; Sadler, A. The role of protein kinase R in the interferon response. *J. Interferon Cytokine Res.* **2010**, *31*, 59–70. [CrossRef] [PubMed]
90. Schulz, O.; Pichlmair, A.; Rehwinkel, J.; Rogers, N.C.; Scheuner, D.; Kato, H.; Takeuchi, O.; Akira, S.; Kaufman, R.J.; Reis e Sousa, C. Protein kinase R contributes to immunity against specific viruses by regulating interferon mRNA integrity. *Cell Host Microbe* **2010**, *7*, 354–361. [CrossRef] [PubMed]
91. Ingelfinger, D.; Arndt-Jovin, D.J.; Lührmann, R.; Achsel, T. The human LSm1–7 proteins colocalize with the mRNA-degrading enzymes Dcp1/2 and Xrnl in distinct cytoplasmic foci. *RNA* **2002**, *8*, 1489–1501. [PubMed]
92. Jain, S.; Parker, R. The discovery and analysis of P bodies. In *Ten Years of Progress in GW/P Body Research*; Chan, L.E.K., Fritzler, J.M., Eds.; Springer: New York, NY, USA, 2013; pp. 23–43.
93. Sheth, U.; Parker, R. Decapping and decay of messenger RNA occur in cytoplasmic processing bodies. *Science* **2003**, *300*, 805–808. [CrossRef] [PubMed]
94. Zheng, D.; Ezzeddine, N.; Chen, C.-Y.A.; Zhu, W.; He, X.; Shyu, A.-B. Deadenylation is prerequisite for P-body formation and mRNA decay in mammalian cells. *J. Cell Biol.* **2008**, *182*, 89–101. [CrossRef] [PubMed]
95. Cougot, N.; Babajko, S.; Séraphin, B. Cytoplasmic foci are sites of mRNA decay in human cells. *J. Cell Biol.* **2004**, *165*, 31–40. [CrossRef] [PubMed]
96. Stalder, L.; Mühlemann, O. Processing bodies are not required for mammalian nonsense-mediated mRNA decay. *RNA* **2009**, *15*, 1265–1273. [CrossRef] [PubMed]
97. Eulalio, A.; Behm-Ansmant, I.; Schweizer, D.; Izaurralde, E. P-body formation is a consequence, not the cause, of RNA-mediated gene silencing. *Mol. Cell. Biol.* **2007**, *27*, 3970–3981. [CrossRef] [PubMed]
98. Liu, J.; Valencia-Sanchez, M.A.; Hannon, G.J.; Parker, R. MicroRNA-dependent localization of targeted mRNAs to mammalian P-bodies. *Nat. Cell Biol.* **2005**, *7*, 719–723. [CrossRef] [PubMed]
99. Pillai, R.S.; Bhattacharyya, S.N.; Artus, C.G.; Zoller, T.; Cougot, N.; Basyuk, E.; Bertrand, E.; Filipowicz, W. Inhibition of Translational Initiation by Let-7 MicroRNA in Human Cells. *Science* **2005**, *309*, 1573–1576. [CrossRef] [PubMed]
100. Sheth, U.; Parker, R. Targeting of aberrant mRNAs to cytoplasmic processing bodies. *Cell* **2006**, *125*, 1095–1109. [CrossRef] [PubMed]
101. Bhattacharyya, S.N.; Habermacher, R.; Martine, U.; Closs, E.I.; Filipowicz, W. Relief of microRNA-mediated translational repression in human cells subjected to stress. *Cell* **2006**, *125*, 1111–1124. [CrossRef] [PubMed]
102. Brengues, M.; Teixeira, D.; Parker, R. Movement of eukaryotic mRNAs between polysomes and cytoplasmic processing bodies. *Science* **2005**, *310*, 486–489. [CrossRef] [PubMed]
103. Andrei, M.A.; Ingelfinger, D.; Heintzmann, R.; Achsel, T.; Rivera-Pomar, R.; Lurhmann, R. A role for eIF4E and eIF4E-transporter in targeting mRNPs to mammalian processing bodies. *RNA* **2005**, *11*, 717–727. [CrossRef] [PubMed]
104. Beisang, D.; Bohjanen, P.R. Perspectives on the ARE as it turns 25 years old. *Wiley Interdiscip. Rev. RNA* **2012**, *3*, 719–731. [CrossRef] [PubMed]
105. Chen, C.-Y.; Gherzi, R.; Ong, S.-E.; Chan, E.L.; Raijmakers, R.; Pruijn, G.J.M.; Stoecklin, G.; Moroni, C.; Mann, M.; Karin, M. AU binding proteins recruit the exosome to degrade ARE-containing mRNAs. *Cell* **2001**, *107*, 451–464. [CrossRef]
106. Mukherjee, D.; Gao, M.; O'Connor, J.P.; Raijmakers, R.; Pruijn, G.; Lutz, C.S.; Wilusz, J. The mammalian exosome mediates the efficient degradation of mRNAs that contain AU-rich elements. *EMBO J.* **2002**, *21*, 165–174. [CrossRef] [PubMed]
107. Gao, M.; Wilusz, C.J.; Peltz, S.W.; Wilusz, J. A novel mRNA-decapping activity in HeLa cytoplasmic extracts is regulated by AU-rich elements. *EMBO J.* **2001**, *20*, 1134–1143. [CrossRef] [PubMed]
108. Stoecklin, G.; Mayo, T.; Anderson, P. ARE-mRNA degradation requires the 5′–3′ decay pathway. *EMBO Rep.* **2006**, *7*, 72–77. [CrossRef] [PubMed]

109. Gherzi, R.; Lee, K.-Y.; Briata, P.; Wegmüller, D.; Moroni, C.; Karin, M.; Chen, C.-Y. A KH domain RNA binding protein, KSRP, promotes ARE-directed mRNA turnover by recruiting the degradation machinery. *Mol. Cell* **2004**, *14*, 571–583. [CrossRef] [PubMed]

110. Lykke-Andersen, J.; Wagner, E. Recruitment and activation of mRNA decay enzymes by two ARE-mediated decay activation domains in the proteins TTP and BRF-1. *Genes Dev.* **2005**, *19*, 351–361. [CrossRef] [PubMed]

111. Brewer, G. An A + U-rich element RNA-binding factor regulates c-myc mRNA stability in vitro. *Mol. Cell. Biol.* **1991**, *11*, 2460–2466. [CrossRef] [PubMed]

112. Brewer, G.; Saccani, S.; Sarkar, S.; Lewis, A.; Pestka, S. Increased interleukin-10 mRNA stability in melanoma cells is associated with decreased levels of A + U-rich element binding factor AUF1. *J. Interferon Cytokine Res.* **2003**, *23*, 553–564. [PubMed]

113. Chen, T.-M.; Hsu, C.-H.; Tsai, S.-J.; Sun, H.S. AUF1 p42 isoform selectively controls both steady-state and PGE2-induced FGF9 mRNA decay. *Nucleic Acids Res.* **2010**, *38*, 8061–8071. [CrossRef] [PubMed]

114. Lin, S.; Wang, W.; Wilson, G.M.; Yang, X.; Brewer, G.; Holbrook, N.J.; Gorospe, M. Down-regulation of cyclin D1 expression by prostaglandin A2 is mediated by enhanced cyclin D1 mRNA turnover. *Mol. Cell. Biol.* **2000**, *20*, 7903–7913. [CrossRef] [PubMed]

115. Sela-Brown, A.; Silver, J.; Brewer, G.; Naveh-Many, T. Identification of AUF1 as a parathyroid hormone mRNA 3′-untranslated region-binding protein that determines parathyroid hormone mRNA stability. *J. Biol. Chem.* **2000**, *275*, 7424–7429. [CrossRef] [PubMed]

116. Xin, H.; Brown, J.A.; Gong, C.; Fan, H.; Brewer, G.; Gnarra, J.R. Association of the von Hippel–Lindau protein with AUF1 and posttranscriptional regulation of VEGFA mRNA. *Mol. Cancer Res.* **2012**, *10*, 108–120. [CrossRef] [PubMed]

117. Liao, B.; Hu, Y.; Brewer, G. Competitive binding of AUF1 and TIAR to MYC mRNA controls its translation. *Nat. Struct. Mol. Biol.* **2007**, *14*, 511–518. [CrossRef] [PubMed]

118. Wagner, B.J.; DeMaria, C.T.; Sun, Y.; Wilson, G.M.; Brewer, G. Structure and genomic organization of the human AUF1 gene: Alternative pre-mRNA splicing generates four protein isoforms. *Genomics* **1998**, *48*, 195–202. [CrossRef] [PubMed]

119. Sarkar, B.; Lu, J.-Y.; Schneider, R.J. Nuclear import and export functions in the different isoforms of the AUF1/heterogeneous nuclear ribonucleoprotein protein family. *J. Biol. Chem.* **2003**, *278*, 20700–20707. [CrossRef] [PubMed]

120. Zucconi, B.E.; Ballin, J.D.; Brewer, B.Y.; Ross, C.R.; Huang, J.; Toth, E.A.; Wilson, G.M. Alternatively expressed domains of AU-rich element RNA-binding protein 1 (AUF1) regulate RNA-binding affinity, RNA-induced protein oligomerization, and the local conformation of bound RNA ligands. *J. Biol. Chem.* **2010**, *285*, 39127–39139. [CrossRef] [PubMed]

121. Arao, Y.; Kuriyama, R.; Kayama, F.; Kato, S. A nuclear matrix-associated factor, SAF-B, interacts with specific isoforms of AUF1/hnRNP D. *Arch. Biochem. Biophys.* **2000**, *380*, 228–236. [CrossRef] [PubMed]

122. Yoon, J.H.; De, S.; Srikantan, S.; Abdelmohsen, K.; Grammatikakis, I.; Kim, J.; Kim, K.M.; Noh, J.H.; White, E.J.; Martindale, J.L.; et al. PAR-CLIP analysis uncovers AUF1 impact on target RNA fate and genome integrity. *Nat. Commun.* **2014**, *5*, 5248. [CrossRef] [PubMed]

123. Fuentes-Pananá, E.M.; Peng, R.; Brewer, G.; Tan, J.; Ling, P.D. Regulation of the Epstein-Barr virus C promoter by AUF1 and the cyclic AMP/protein kinase A signaling pathway. *J. Virol.* **2000**, *74*, 8166–8175. [CrossRef] [PubMed]

124. Lee, N.; Pimienta, G.; Steitz, J.A. AUF1/hnRNP D is a novel protein partner of the EBER1 noncoding RNA of Epstein-Barr virus. *RNA* **2012**, *18*, 2073–2082. [CrossRef] [PubMed]

125. Lund, N.; Milev, M.P.; Wong, R.; Sanmuganantham, T.; Woolaway, K.; Chabot, B.; Abou Elela, S.; Mouland, A.J.; Cochrane, A. Differential effects of hnRNP D/AUF1 isoforms on HIV-1 gene expression. *Nucleic Acids Res.* **2012**, *40*, 3663–3675. [CrossRef] [PubMed]

126. Paek, K.Y.; Kim, C.S.; Park, S.M.; Kim, J.H.; Jang, S.K. RNA-binding protein hnRNP D modulates internal ribosome entry site-dependent translation of hepatitis C virus RNA. *J. Virol.* **2008**, *82*, 12082–12093. [CrossRef] [PubMed]

127. Friedrich, S.; Schmidt, T.; Geissler, R.; Lilie, H.; Chabierski, S.; Ulbert, S.; Liebert, U.G.; Golbik, R.P.; Behrens, S.-E. AUF1 p45 promotes West Nile virus replication by an RNA chaperone activity that supports cyclization of the viral genome. *J. Virol.* **2014**, *88*, 11586–11599. [CrossRef] [PubMed]

128. Friedrich, S.; Schmidt, T.; Schierhorn, A.; Lilie, H.; Szczepankiewicz, G.; Bergs, S.; Liebert, U.G.; Golbik, R.P.; Behrens, S.E. Arginine methylation enhances the RNA chaperone activity of the West Nile virus host factor AUF1 p45. *RNA* **2016**, *22*, 1574–1591. [CrossRef] [PubMed]

129. Cathcart, A.L.; Rozovics, J.M.; Semler, B.L. Cellular mRNA decay protein AUF1 negatively regulates enterovirus and human rhinovirus infections. *J. Virol.* **2013**, *87*, 10423–10434. [CrossRef] [PubMed]

130. Cathcart, A.L.; Semler, B.L. Differential restriction patterns of mRNA decay factor AUF1 during picornavirus infections. *J. Gen. Virol.* **2014**, *95*, 1488–1492. [CrossRef] [PubMed]

131. Wong, J.; Si, X.; Angeles, A.; Zhang, J.; Shi, J.; Fung, G.; Jagdeo, J.; Wang, T.; Zhong, Z.; Jan, E.; et al. Cytoplasmic redistribution and cleavage of AUF1 during coxsackievirus infection enhance the stability of its viral genome. *FASEB J.* **2013**, *27*, 2777–2787. [CrossRef] [PubMed]

132. Lin, J.-Y.; Li, M.-L.; Brewer, G. mRNA decay factor AUF1 binds the internal ribosomal entry site of enterovirus 71 and inhibits virus replication. *PLoS ONE* **2014**, *9*, e103827. [CrossRef] [PubMed]

133. Rozovics, J.M.; Chase, A.J.; Cathcart, A.L.; Chou, W.; Gershon, P.D.; Palusa, S.; Wilusz, J.; Semler, B.L. Picornavirus modification of a host mRNA decay protein. *mBio* **2012**, *3*, e00431-12. [CrossRef] [PubMed]

134. Spurrell, J.C.L.; Wiehler, S.; Zaheer, R.S.; Sanders, S.P.; Proud, D. Human airway epithelial cells produce IP-10 (CXCL10) in vitro and in vivo upon rhinovirus infection. *Am. J. Physiol. Lung Cell. Mol. Physiol.* **2005**, *289*, L85–L95. [CrossRef] [PubMed]

135. Lin, J.-Y.; Shih, S.-R.; Pan, M.; Li, C.; Lue, C.-F.; Stollar, V.; Li, M.-L. hnRNP A1 interacts with the 5′ untranslated regions of enterovirus 71 and Sindbis virus RNA and is required for viral replication. *J. Virol.* **2009**, *83*, 6106–6114. [CrossRef] [PubMed]

136. Levengood, J.D.; Tolbert, M.; Li, M.-L.; Tolbert, B.S. High-affinity interaction of hnRNP A1 with conserved RNA structural elements is required for translation and replication of enterovirus 71. *RNA Biol.* **2013**, *10*, 1136–1145. [CrossRef] [PubMed]

137. Cammas, A.; Pileur, F.; Bonnal, S.; Lewis, S.M.; Lévêque, N.; Holcik, M.; Vagner, S. Cytoplasmic relocalization of heterogeneous nuclear ribonucleoprotein A1 controls translation initiation of specific mRNAs. *Mol. Biol. Cell* **2007**, *18*, 5048–5059. [CrossRef] [PubMed]

138. Min, H.; Turck, C.W.; Nikolic, J.M.; Black, D.L. A new regulatory protein, KSRP, mediates exon inclusion through an intronic splicing enhancer. *Genes Dev.* **1997**, *11*, 1023–1036. [CrossRef] [PubMed]

139. Davis-Smyth, T.; Duncan, R.C.; Zheng, T.; Michelotti, G.; Levens, D. The far upstream element-binding proteins comprise an ancient family of single-strand DNA-binding transactivators. *J. Biol. Chem.* **1996**, *271*, 31679–31687. [CrossRef] [PubMed]

140. Trabucchi, M.; Briata, P.; Garcia-Mayoral, M.; Haase, A.D.; Filipowicz, W.; Ramos, A.; Gherzi, R.; Rosenfeld, M.G. The RNA-binding protein KSRP promotes the biogenesis of a subset of microRNAs. *Nature* **2009**, *459*, 1010–1014. [CrossRef] [PubMed]

141. Lin, J.-Y.; Li, M.-L.; Shih, S.-R. Far upstream element binding protein 2 interacts with enterovirus 71 internal ribosomal entry site and negatively regulates viral translation. *Nucleic Acids Res.* **2009**, *37*, 47–59. [CrossRef] [PubMed]

142. Chen, L.-L.; Kung, Y.-A.; Weng, K.-F.; Lin, J.-Y.; Horng, J.-T.; Shih, S.-R. Enterovirus 71 infection cleaves a negative regulator for viral internal ribosomal entry site-driven translation. *J. Virol.* **2013**, *87*, 3828–3838. [CrossRef] [PubMed]

143. Jean-Philippe, J.; Paz, S.; Caputi, M. hnRNP A1: The swiss army knife of gene expression. *Int. J. Mol. Sci.* **2013**, *14*, 18999–19024. [CrossRef] [PubMed]

144. Zhao, T.T.; Graber, T.E.; Jordan, L.E.; Cloutier, M.; Lewis, S.M.; Goulet, I.; Cote, J.; Holcik, M. hnRNP A1 regulates UV-induced NF-[kappa]B signalling through destabilization of cIAP1 mRNA. *Cell Death Differ.* **2008**, *16*, 244–252. [CrossRef] [PubMed]

145. Geissler, R.; Simkin, A.; Floss, D.; Patel, R.; Fogarty, E.A.; Scheller, J.; Grimson, A. A widespread sequence-specific mRNA decay pathway mediated by hnRNPs A1 and A2/B1. *Genes Dev.* **2016**, *30*, 1070–1085. [CrossRef] [PubMed]

146. Bonnal, S.; Pileur, F.; Orsini, C.; Parker, F.; Pujol, F.; Prats, A.-C.; Vagner, S. Heterogeneous nuclear ribonucleoprotein A1 is a novel internal ribosome entry site trans-acting factor that modulates alternative initiation of translation of the fibroblast growth factor 2 mRNA. *J. Biol. Chem.* **2005**, *280*, 4144–4153. [CrossRef] [PubMed]

147. Kunze, M.M.; Benz, F.; Brauß, T.F.; Lampe, S.; Weigand, J.E.; Braun, J.; Richter, F.M.; Wittig, I.; Brüne, B.; Schmid, T. sST2 translation is regulated by FGF2 via an hnRNP A1-mediated IRES-dependent mechanism. *BBA Gene Regul. Mech.* **2016**, *1859*, 848–859. [CrossRef] [PubMed]

148. Iwakawa, H.-o.; Tomari, Y. The functions of microRNAs: MRNA decay and translational repression. *Trends Cell Biol.* **2015**, *25*, 651–665. [CrossRef] [PubMed]

149. Lee, Y.; Ahn, C.; Han, J.; Choi, H.; Kim, J.; Yim, J.; Lee, J.; Provost, P.; Radmark, O.; Kim, S.; Kim, V.N. The nuclear RNase III Drosha initiates microRNA processing. *Nature* **2003**, *425*, 415–419. [CrossRef] [PubMed]

150. Bernstein, E.; Caudy, A.A.; Hammond, S.M.; Hannon, G.J. Role for a bidentate ribonuclease in the initiation step of RNA interference. *Nature* **2001**, *409*, 363–366. [CrossRef] [PubMed]

151. Azlan, A.; Dzaki, N.; Azzam, G. Argonaute: The executor of small RNA function. *J. Genet. Genom.* **2016**, *43*, 481–494. [CrossRef] [PubMed]

152. Liu, J.; Carmell, M.A.; Rivas, F.V.; Marsden, C.G.; Thomson, J.M.; Song, J.-J.; Hammond, S.M.; Joshua-Tor, L.; Hannon, G.J. Argonaute2 is the catalytic engine of mammalian RNAi. *Science* **2004**, *305*, 1437–1441. [CrossRef] [PubMed]

153. Maillard, P.V.; Ciaudo, C.; Marchais, A.; Li, Y.; Jay, F.; Ding, S.W.; Voinnet, O. Antiviral RNA interference in mammalian cells. *Science* **2013**, *342*, 235–238. [CrossRef] [PubMed]

154. Shi, J.; Sun, J.; Wang, B.; Wu, M.; Zhang, J.; Duan, Z.; Wang, H.; Hu, N.; Hu, Y. Novel microRNA-like viral small regulatory RNAs arising during human hepatitis A virus infection. *FASEB J.* **2014**, *28*, 4381–4393. [CrossRef] [PubMed]

155. Weng, K.-F.; Hung, C.-T.; Hsieh, P.-T.; Li, M.-L.; Chen, G.-W.; Kung, Y.-A.; Huang, P.-N.; Kuo, R.-L.; Chen, L.-L.; Lin, J.-Y.; et al. A cytoplasmic RNA virus generates functional viral small RNAs and regulates viral IRES activity in mammalian cells. *Nucleic Acids Res.* **2014**, *42*, 12789–12805. [CrossRef] [PubMed]

MDPI

Review

Interactions between the HIV-1 Unspliced mRNA and Host mRNA Decay Machineries

Daniela Toro-Ascuy *, Bárbara Rojas-Araya, Fernando Valiente-Echeverría and Ricardo Soto-Rifo *

Molecular and Cellular Virology Laboratory, Virology Program, Institute of Biomedical Sciences, Faculty of Medicine, Universidad of Chile, Independencia 834100, Santiago, Chile; barbara.rojas.araya@gmail.com (B.R.-A.); fvaliente@med.uchile.cl (F.V.-E.)

* Correspondence: daniela.toroascuy@gmail.com (D.T.-A.); rsotorifo@med.uchile.cl (R.S.-R.); Tel.: +56-2-978-96-16 (D.T.-A.); +56-2-978-68-69 (R.S.-R.)

Academic Editors: J. Robert Hogg and Karen L. Beemon
Received: 1 October 2016; Accepted: 14 November 2016; Published: 23 November 2016

Abstract: The human immunodeficiency virus type-1 (HIV-1) unspliced transcript is used both as mRNA for the synthesis of structural proteins and as the packaged genome. Given the presence of retained introns and instability AU-rich sequences, this viral transcript is normally retained and degraded in the nucleus of host cells unless the viral protein REV is present. As such, the stability of the HIV-1 unspliced mRNA must be particularly controlled in the nucleus and the cytoplasm in order to ensure proper levels of this viral mRNA for translation and viral particle formation. During its journey, the HIV-1 unspliced mRNA assembles into highly specific messenger ribonucleoproteins (mRNPs) containing many different host proteins, amongst which are well-known regulators of cytoplasmic mRNA decay pathways such as up-frameshift suppressor 1 homolog (UPF1), Staufen double-stranded RNA binding protein 1/2 (STAU1/2), or components of miRNA-induced silencing complex (miRISC) and processing bodies (PBs). More recently, the HIV-1 unspliced mRNA was shown to contain N^6-methyladenosine (m^6A), allowing the recruitment of YTH N^6-methyladenosine RNA binding protein 2 (YTHDF2), an m^6A reader host protein involved in mRNA decay. Interestingly, these host proteins involved in mRNA decay were shown to play positive roles in viral gene expression and viral particle assembly, suggesting that HIV-1 interacts with mRNA decay components to successfully accomplish viral replication. This review summarizes the state of the art in terms of the interactions between HIV-1 unspliced mRNA and components of different host mRNA decay machineries.

Keywords: HIV-1 unspliced mRNA; mRNA decay; REV; UPF1; Staufen; m^6A; YTHDF2

1. Introduction

Eukaryotic cells employ quality control mechanisms to ensure that each step of mRNA metabolism, from transcription to translation and decay, is properly executed in space and time and the genetic code is correctly expressed. mRNA surveillance and decay pathways are responsible for recognizing aberrant mRNAs that arise due to errors in the DNA template or by misprocesses occurring during mRNA biogenesis [1]. As such, efficient and accurate gene expression is ensured by mechanisms that degrade mRNAs in the nucleus as a response to defects in transcription elongation [2], splicing [3], 3'-end formation [3], and nuclear export [4]. Following export to the cytoplasm, nonsense-mediated decay (NMD), a ribosome-coupled quality control mechanism, induces degradation of mRNAs that contain premature termination codons [5]. Other translation-dependent mechanisms of mRNA degradation are the no-go decay (NGD) pathway, which leads to endonucleolytic cleavage of mRNAs containing strong stalls in translation elongation [6,7], and the non-stop decay (NSD), which corresponds to a quality control mechanism that detects mRNA molecules lacking a stop codon [8].

In addition, some mRNAs harbor specific *cis* signals including miRNA target sites, AU-rich elements, and methylated adenosines (N^6-methyladenosine or m^6A), which have been involved in the control of mRNA stability [9–14].

Upon viral infection, host cells mount an antiviral stress response in order to create a hostile environment for viral replication. This cellular response usually involves the shut-off of protein synthesis and the concomitant assembly of RNA granules such as stress granules (SGs), which correspond to sites of mRNA triage and PBs, which contain the mRNA degradation machinery [15]. Given the fact that most positive single-stranded RNA viruses including poliovirus (PV), hepatitis C virus (HCV), and human immunodeficiency virus (HIV) use the same molecule first as mRNA and then as the packaged genome, it is not surprising that these viruses have evolved different mechanisms aimed at modulating the assembly of different RNA granules and counteracting mRNA decay machineries [15]. Indeed, there is increasing evidence indicating that these viruses are able to interact with and/or modify the cellular factors implicated in mRNA quality control mechanisms during different steps of their replication cycle [15].

This review summarizes the state-of-the-art in terms of the interactions between the HIV-1 unspliced mRNA and proteins with cellular factors involved in different mRNA decay pathways. We also discuss the potential strategies the virus has evolved to divert some of these mRNA degradation pathways or their components and to favor viral gene expression and replication.

2. An Overview on Human Immunodeficiency Virus Type-1 (HIV-1) Gene Expression

HIV-1 is the prototype member of the *Lentivirus* genus of the *Retroviridae* family and the etiologic agent of the acquired immunodeficiency syndrome (AIDS). The HIV-1 genome consists of a 9 kb single-stranded RNA molecule carrying nine open reading frames that give rise to 15 viral proteins [16]. Moreover, transcription from the 3′-long terminal repeat (LTR) promoter gives rise to an additional protein named antisense protein (ASP), which seems to be important for HIV spread and pandemics [17]. Once integrated into a host chromosome, HIV-1 gene expression is regulated at the transcriptional and post-transcriptional levels by viral proteins TAT and REV, which are supported by several host proteins [18] (Figure 1). Host RNA polymerase II drives the synthesis of the full-length 9 kb mRNA, which is identical to the genomic RNA (gRNA) present within viral particles. Early during viral gene expression, the full-length transcript recruits the host mRNA processing machinery and undergoes alternative splicing, generating a subset of fully spliced (2 kb) and partially spliced (4 kb) transcripts, which in addition to the unspliced mRNA, are responsible for the synthesis of all viral proteins [19–22]. Fully spliced transcripts (used for the synthesis of TAT, REV, and NEF) follow the same classical gene expression pathway as cellular mRNAs, in which the rates of nuclear export and translation are highly stimulated by the splicing-dependent recruitment of nuclear proteins including the mRNA export factor NXF1 and the exon junction complex (EJC), amongst others (Figure 1). In sharp contrast, partially spliced transcripts (coding for ENV, VPU, VIF, and VPR) [19,20] and the unspliced mRNA (used for GAG and GAG–POL synthesis) [20] are not able to follow the classical mRNA nuclear export pathway due to the presence of introns and thus rely on an alternative mechanism of nuclear export in order to evade NXF1-associated quality controls.

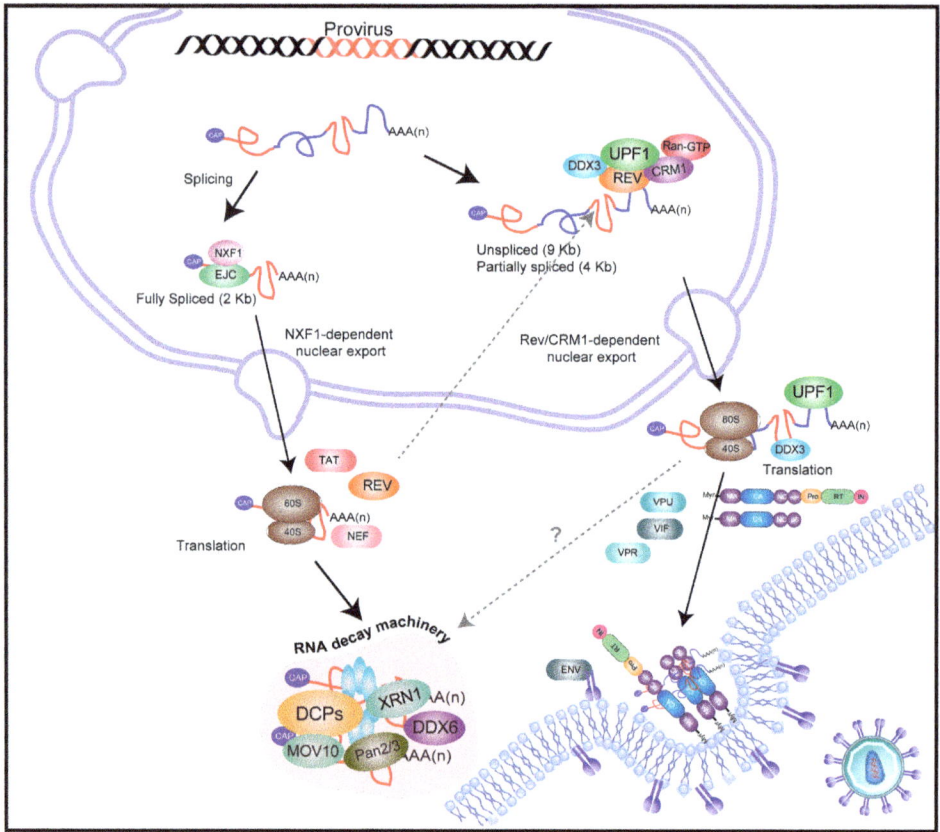

Figure 1. Post-transcriptional control of gene expression in HIV-1. Upon RNA polymerase II-driven transcription, the capped and polyadenylated 9 kb full-length mRNA undergoes alternative splicing in order to generate the 2 kb fully spliced and the 4 kb partially spliced (omitted for simplicity) transcripts. Fully spliced transcripts follow the canonical pathway for mRNA metabolism, in which nuclear export and translation are ensured by the splicing-dependent recruitment of nuclear factors such as the exon junction complex (EJC) and the mRNA nuclear export factor NXF1. Once in the cytoplasm, fully spliced mRNA recruits the host translational machinery in order to synthesize viral proteins TAT, REV, and NEF and upon several rounds of translation they are degraded by the host RNA decay machinery. The viral protein REV enters the nucleus, allowing the accumulation of the 9 kb unspliced mRNA and its subsequent nuclear export through the chromosomal maintenance 1 (CRM1)-dependent pathway. This alternative nuclear export pathway allows the unspliced mRNA to evade surveillance and quality control mechanisms associated with the canonical nuclear export pathway. During its journey to the cytoplasm, the unspliced mRNA recruits several host proteins such as up-frameshift suppressor 1 homolog (UPF1) and the DEAD (Asp-Glu-Ala-Asp) box 3- (DDX3) RNA helicase that will ensure an efficient association with the host translational machinery in order to synthesize the major structural proteins GAG and GAG–POL. In contrast to fully spliced transcripts, the unspliced mRNA does not undergo mRNA turnover as it is used as the viral genome incorporated into viral particles. CA, capsid protein; DCP, decapping enzyme; IN, integrase; MA, matrix protein; MOV10; Moloney leukemia virus 10; Myr, N-terminally myristoylated; NC, nucleocapsid protein; Pan2/3, PAB-dependent poly(A)-specific ribonuclease; p6, p6 protein; Ran-GTP, ras-related nuclear protein GTP; RT, reverse transcriptase; XRN1, 5′-3′ Exoribonuclease 1.

3. The Full-Length and Partially Spliced mRNAs of HIV-1 Evade Nuclear Surveillance and Quality Control Cellular Mechanisms

As mentioned above, fully spliced transcripts follow the classical gene expression pathway employed by cellular mRNAs and are expected to undergo nucleoporin Tpr-mediated surveillance at the nuclear pore complex [23,24]. In sharp contrast, partially spliced transcripts and the unspliced mRNA are not able to follow the classical mRNA nuclear export pathway due to the presence of introns, which are recognized by this NXF1-associated mRNA surveillance mechanism, which induces nuclear retention and degradation of unprocessed transcripts [23,24]. However, the virus has evolved the REV protein, which binds to a specific RNA structure (the REV-responsive element or RRE) present exclusively within these intron-containing transcripts and to the host karyopherin chromosomal maintenance 1 (CRM1) [25–27]. In addition to the leucine-rich nuclear export signal (NES) that allows its association with CRM1, REV also possess a nuclear localization signal (NLS) that is recognized by importins-α/β, allowing for shuttling between the nucleus and cytoplasm through nuclear pore complexes (NPCs) [28–30].

As mentioned above, the REV protein was shown to be required for the transport of the unspliced and partially spliced mRNAs from the nucleus to the cytoplasm by a non-canonical mRNA export pathway [18,31]. Indeed, these intron-containing viral mRNAs are retained and degraded in the nucleus in the absence of REV because of their incomplete splicing [32]. At the molecular level, REV binds and oligomerizes along the 351nt RRE located within the *env* gene and thus is present exclusively in all underspliced viral transcripts [33,34]. Once synthesized, REV is imported and accumulates in the nucleus, where it binds to RRE-containing transcripts to promote their export via CRM1 [35]. It is noteworthy that the main function of CRM1 is related to the nuclear export of NES-containing proteins and small RNAs, being only infrequently involved in cellular mRNA export [35]. As a consequence, intron-containing viral transcripts avoid the mRNA surveillance and quality control mechanisms associated with the canonical nuclear export pathway and accumulate in the cytoplasm for translation [35] (Figure 1).

In addition to CRM1, a large number of cellular proteins have been shown to influence REV's functions in the nuclear export of unspliced and partially spliced viral mRNAs [36]. These host factors include Matrin 3, host nuclear matrix protein (MATR3), an important host factor required to stabilize the viral RNA through its interaction with REV/RRE [37], REV-interacting protein (RIP)–REV/Rex activation domain-binding protein (RAB) [38,39], eukaryotic translation initiation factor 5A (eIF5A), Src-associated substrate in mitosis of 68 kDa (Sam68), and RNA helicases such as DEAD (Asp-Glu-Ala-Asp) box 3 (DDX3) [40], DDX1 [41], and RNA helicase A (RHA) [42], amongst others. Whether these proteins contribute to the stabilization of the viral nuclear export of mRNP or whether they determine the cytoplasmic fate of the viral mRNAs once exported (translation or packaging) requires further investigation.

4. REV Stabilizes RNA Instability Elements Present within the HIV-1 Unspliced mRNA

In contrast to cytoplasmic mRNA quality control pathways such as NMD, nuclear mRNA turnover is less understood. Interestingly, HIV-1 intron-containing mRNAs undergo nuclear downregulation as they are further spliced to completion or degraded in the absence of REV [32,43–45]. In the late 1980s, Pavlakis′s group designed an experimental setting aimed to identify inhibitory sequences present within the HIV-1 genome and to further study their function on viral gene expression [46–48]. They identified and characterized an inhibitory sequence in the HIV-1 *gag* gene that was named INS-1 [46]. They showed that the INS-1 element does not contain any functional splice site and acts in *cis* by lowering steady-state mRNA levels. Thus, INS-1 appeared to function at the level of mRNA stability [46]. The authors suggested that the inhibitory effect of INS-1 could be overcome by the REV–RRE interaction, demonstrating that this sequence present within the *gag* gene was important for REV-regulated viral gene expression [46]. Subsequently, the same group and others described more inhibitory sequences present in other genomic locations such as the *gag*/*pol* intersection (IN) [45], and within the *pol* (*cis*-repressive sequences or CRS) [49] and *env* [48] genes. These elements were shown to interfere with viral gene expression by impairing mRNA stability, nucleocytoplasmic transport, and cap-dependent translation initiation [46,50]. Interestingly, REV counteracted the defects exerted by these mRNA instability elements, allowing efficient viral gene expression [51].

Later on, Schneider and colleagues observed that most of the regions linked to instability (INS) contained high AU contents. Interestingly, while all REV-dependent mRNAs have unusually high AU contents, the AU content of fully spliced mRNA species is much lower [52]. Indeed, the AU contents within INS regions vary between 46% and 92% (with the average AU content in cellular mRNA being around 50%). However, it has been observed that particularly unstable cellular mRNAs such as c-myc, c-fos, c-myb, granulocyte-macrophage colony-stimulating factor, or mRNAs coding for cytokines share unusually high AU contents, which are involved in the instability and rapid degradation of these transcripts [53]. Interestingly, some of the viral INS elements contain the AUUUA pentanucleotide, which corresponds to a signal (AU-rich element or ARE) known to trigger an mRNA decay pathway known as ARE-mediated decay [53–56]. It is important to note that REV is unable to export underspliced mRNAs that do not contain a functional INS and hence it was proposed that these instability regions are an integral component of REV regulation [50]. Several mRNA-binding proteins, including polypyrimidine tract-binding protein 1 (PTB)/heterogeneous nuclear ribonucleoprotein I (hnRNP I) [57], heterogeneous nuclear ribonucleoprotein A1 (hnRNPA1) [58], and poly(A) binding protein cytoplasmic 1 (PABPC1) [47], were shown to specifically bind to such elements in vitro. It has been suggested that these INS-binding factors may avoid the recognition of the unspliced mRNA by the splicing machinery and promote their association with REV, thus enabling their export and expression. However, the precise molecular mechanisms by which INS and INS-binding factors acts on HIV-1 gene expression are still uncertain.

Subsequently, Zolotukhin and coworkers showed that the INS region present within the HIV-1 gag mRNA was bound by the heterodimeric transcription/splicing factor p54nrb/polypyrimidine tract-binding protein-associated splicing factor (PSF) [51]. By performing functional assays, the authors showed that PSF subunits act at the post-transcriptional level via INS in order to inhibit gag mRNA expression [51]. The authors proposed that p54nrb and PSF were host factors mediating INS function through a probably novel mRNA regulatory pathway regulating HIV-1 unspliced mRNA expression. However, a recent report showed that PSF was a positive component of the REV-mediated axis, whose contribution was to ensure a pool of underspliced mRNAs available for REV activity [37].

More recently, Valiente-Echeverría and colleagues reported the inhibitory effect of INS-1 on HIV-1 internal ribosome entry site (IRES)-mediated translation initiation. By using heterologous bicistronic mRNAs and both in vitro and cell-based assays, the authors showed that ectopic expression of REV and hnRNPA1 partially rescued the inhibition of INS-1 on translation [59].

It is clear that INS and other instability sequences are *cis* elements important for HIV-1 RNA and protein homeostasis and that the viral protein REV is involved in such regulation. However, it is still unclear what molecular mechanisms are in play and thus further investigation is needed to better understand this regulation.

5. HIV-1 Recruits Factors Involved in Cytoplasmic mRNA Decay Pathways

Cytoplasmic quality control pathways include NMD [60,61], NSD [62], and NGD [7], each of which depends on mRNA translation. The main function of these cytoplasmic mRNA decay pathways is to ensure the fidelity of gene expression [63,64]. As mentioned above, other pathways of cytoplasmic mRNA decay are conditionally used to regulate gene expression of specific mRNA targets containing specific *cis*-acting elements such as ARE-mediated decay [11], miRNA-mediated mRNA decay [65], or Staufen STAU-mediated mRNA decay (SMD) [61,63,66].

5.1. Nonsense-Mediated Decay

Nonsense-mediated mRNA decay is a quality control mechanism playing an important role in the degradation of mRNAs harboring premature termination codons (PTCs), thus avoiding the synthesis of truncated proteins that could be deleterious for the cell [67,68] (Figure 2). The activation of NMD depends on the conserved function of the UPF proteins UPF1, UPF2, and UPF3 [69,70]. UPF1 has a RNA helicase activity essential for NMD, while UPF2 serves as a bridge between UPF1 and UPF3. UPF3 interacts with the mRNA-bound EJC components eukaryotic translation initiation factor 4A3 (eIF4AIII), Y14, and Mago homolog (MAGOH) [71,72]. It has been estimated that around 5% to 20% of cellular mRNAs are NMD substrates, although it has not been established that every potential NMD substrate undergoes NMD-mediated degradation [71]. As viral mRNAs associate with the host machineries for processing, nuclear export, and translation, the question of how NMD affects viral mRNAs arises [72]. In this regard, various reports have shown that some RNA viruses have developed strategies to directly inhibit NMD and thus avoid this cytoplasmic mRNA degradation mechanism [73]. Several reports have shown that HIV-1 recruits the major NMD factor UPF1 to viral mRNPs containing the unspliced mRNA. In a pioneering report, Mouland´s group reported that UPF1 played unexpected roles in HIV-1 unspliced mRNA metabolism by promoting both nuclear export and translation [74]. In this work, the authors showed that UPF1 knockdown resulted in a strong decrease in HIV-1 unspliced mRNA levels and GAG expression (Figure 2a). Consistent with a positive effect on gene expression, the authors observed that overexpression of UPF1 resulted in enhanced levels of both the unspliced mRNA and GAG [74]. By using different mutants, the authors also demonstrated that the role of UPF1 in HIV-1 gene expression could be separated from its functions in NMD [74]. Interestingly, Hogg and Goff reported that UPF1 was able to sense reporter RNAs bearing the HIV-1 3'-untranslated region (UTR) and trigger mRNA decay in a 3'-UTR length-dependent manner [75]. However, further analyses using the whole virus confirmed that UPF1 indeed increases the levels of viral mRNA and the expression of GAG protein during the replication cycle [76,77]. Indeed, Mouland's group provided further insights into the molecular mechanism by which UPF1 regulates the fate of the unspliced mRNA [76]. As such, Ajamian and colleagues demonstrated that UPF1 promotes HIV-1 unspliced mRNA gene expression by forming a specific complex with REV, CRM1, and the DEAD-box RNA helicase DDX3 [76] (Figure 2a). Interestingly, the authors also demonstrated that UPF2 was excluded from these specific UPF1/HIV-1 mRNP [76]. Protein–protein docking suggested that HIV-1 REV could bind UPF1 in a region that overlaps the UPF2 binding site. These in silico tests could explain the exclusion of UPF2, which acts as a negative regulator of gene expression from the unspliced mRNA [76]. The positive effects of UPF1 on HIV RNA metabolism reported in the context of a full replication cycle rather than with reporter RNAs support a model in which this host protein is not a decay-inducing factor for the HIV RNA [78].

Besides its functions on the post-transcriptional regulation of the unspliced mRNA, UPF1 was also shown to be critical for early events of the HIV-1 replication cycle. As such, Serquiña and coworkers reported that UPF1 knockdown or the ectopic expression of ATPase activity mutants resulted in reduced viral entry and reverse transcription (RT) [77]. Interestingly, the authors demonstrated that UPF1 was incorporated into viral particles through specific interactions with the nucleocapsid (NC) domain of GAG [77].

Interestingly, RNA editing enzymes such as the APOBEC3 cytidine deaminases may promote the generation of PTCs along the HIV RNA and thus its subsequent degradation through NMD. Indeed, in the absence of the viral protein Vif, apolipoprotein-B-mRNA-editing enzyme catalytic polypeptide-like 3G (APOBEC3G) generates as much as 20% of dC to dU changes by deamination of the minus-strand during the reverse transcription process [78,79]. However, it is still unknown whether APOBEC3-mediated hypermutations elicit NMD of the modified viral RNA.

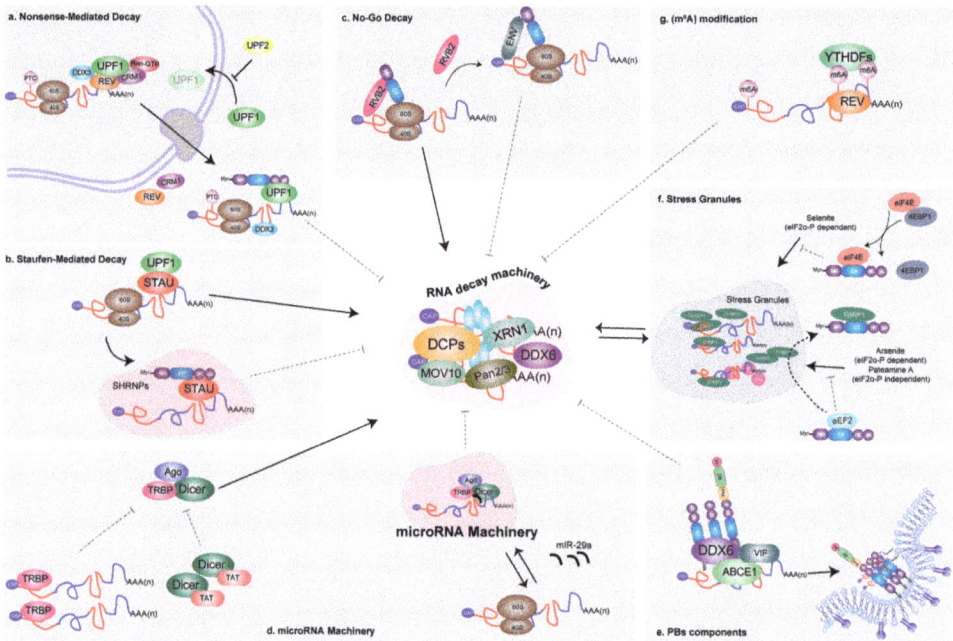

Figure 2. The HIV-1 unspliced mRNA has been shown to recruit components of different mRNA decay pathways including: (**a**) UPF1 (NMD, nonsense-mediated decay); (**b**) STAU1/2 (SMD, STAU-mediated mRNA decay); (**c**) RuvB-like 2 (RVB2) (NGD, No-Go decay); (**d**) HIV-1 *trans*-activating response (TAR) RNA-binding protein (TRBP) and Argonaute (Ago) (microRNA Machinery); (**e**) DDX6 (PBs, processing bodies); (**f**) eukaryotic elongation factor 2 (eEF2) and GTPase activating protein (GAP) SH3 domain-binding protein 1 (G3BP1) (SGs, stress granules); and (**g**) YTHDF2 (N^6-methyladenosine (m^{6}A)-dependent mRNA decay). Interestingly, most of these associations have been demonstrated to be beneficial for viral replication, suggesting that HIV-1 has evolved mechanisms to interact with these host factors in order to divert them from their functions in mRNA decay.

Together, these results strongly indicate that the mRNA decay factor UPF1 is critical in determining the fate of the unspliced mRNA but also during the early steps of viral replication. The molecular determinants that interfere with the UPF1-mediated RNA decay pathway that senses the length of the HIV-1 3′-UTR in the context of a replication cycle are still unknown. Thus, further studies are required to determine whether UPF1 recruitment to the HIV-1 mRNPs interferes with RNA decay-promoting activities.

5.2. Staufen-Mediated mRNA Decay

Staufen (STAU) proteins are involved in multiple post-transcriptional regulatory processes, such as the regulation of mRNA transport and the activation of localized mRNA translation in neurons [80,81], as well as the binding to sequences present within the 3′-UTR of mRNAs [82,83]. Likewise, it has also been proposed that STAU can mediate the degradation of mRNA through the interaction with UPF1 in a process known as STAU-mediated mRNA decay [84].

STAU-mediated mRNA decay is an mRNA degradation process occurring in mammalian cells that is mediated by the binding of Staufen to a STAU1-binding site (SBS) present within the 3′-UTR of target mRNAs [84]. During this process, STAU1 recognizes dsRNA structures formed within the 3′-UTR of target mRNAs but also by an intermolecular association between the 3′-UTR of a target mRNA and complementary Alu elements present in long-noncoding RNA (lncRNA). The STAU1 paralog, STAU2, has also been reported to mediate SMD and both STAU proteins interact with UPF1, which is a key factor required for SMD [84]. Several reports have demonstrated that HIV-1 unspliced mRNA and GAG protein recruit STAU1 to form a specific viral mRNP (Figure 2b). By using a reporter gene harboring the *trans*-activating response region (TAR) at the 5′-end, Dugré-Brisson and colleagues presented evidence showing that STAU1 interacts with TAR, facilitating translation [85]. It was suggested that STAU1 might facilitate the nucleo–cytoplasmic transport of transcripts containing TAR and contribute to their interaction with the host translational machinery [85]. However, the mechanism by which this occurs has not yet been described. In addition, Chatel-Chaix and colleagues showed that STAU1 is an integral component of an intracellular HIV-1 ribonucleoprotein complex containing GAG [86–88]. Furthermore, the authors demonstrated that STAU1 interacts specifically with the NC domain of GAG in an RNA-independent manner [86]. The authors also showed that the HIV-1 unspliced mRNA co-immunoprecipitates together with STAU1, indicating that the viral mRNA is bound by STAU1 and the specific knockdown of STAU1 resulted in a significant reduction in viral infectivity [86]. Later, the same research group reported the assembly of a novel STAU1 RNP whose formation was dependent on HIV-1. STAU1, unspliced mRNA, and GAG colocalize in these STAU1 HIV-1-dependent RNPs (SHRNPs), the size of which depends on existing STAU1 levels in cells [89] (Figure 2b).

Other studies have determined that the human protein STAU2 promotes the export of HIV-1 mRNAs containing an RRE. This effect was shown to occur through the mRNA-independent interaction between REV and STAU2 [90]. Disruption of the REV–STAU2 interaction interferes with viral replication, indicating that recruitment of STAU2 to the RRE (which is located at the 3′-UTR of the unspliced mRNA) is critical for the HIV life cycle.

Together, these data strongly suggest that HIV-1 interacts with STAU proteins to form specific viral mRNPs that are required for efficient gene expression, trafficking, and viral particle assembly. It is unclear whether the recruitment of STAU proteins is related to a virally induced inhibition of SMD.

5.3. No-Go Decay

Recent findings suggest that HIV-1 may exploit the NGD pathway to fine-tune its own gene expression and ensure production of infectious virions. As such, Mu and colleagues showed that RuvB-like 2 (RVB2) inhibits HIV-1 GAG expression and that this inhibitory activity is antagonized by the viral ENV protein [91] (Figure 2c). These authors found that the HIV-1 unspliced mRNA is susceptible to NGD through a mechanism dependent on the translation of the matrix domain (MA) of GAG [91]. The authors also demonstrated that the RVB2 ATPase interacts with the HIV-1 5′-UTR and nascent MA peptides, impeding further translation of GAG or GAG–POL (Figure 2c). Thus, it was proposed that this mechanism mediated by RVB2 allows a balance between GAG and ENV by regulating the relative expression levels of these structural viral proteins necessary for efficient production of infectious viral particles [91]. Thus, by using the NGD mechanism, HIV-1 exploits a host RNA quality control pathway to maximize the quality of viral particles [91].

Together, these studies strongly suggest that HIV-1 proteins and/or RNA recruit factors involved in the translation-dependent degradation of cellular mRNAs such as UPF1 and STAU1 in order to ensure efficient viral replication. However, whether such interactions interfere with the mRNA degradation processes needs to be further investigated. Thus, studies aimed at identifying other factors that are involved in these pathways would be useful to clarify how the virus evades or interferes with mRNA quality control mechanisms.

6. Relationship between HIV and the Cellular microRNA Machinery and Processing Bodies (PBs) Components

RNA silencing is a mechanism for regulation of gene expression involving small non-coding RNA [92], as well as an innate host cell defense mechanism against viruses [93]. miRNA biogenesis begins with the RNA polymerase II-mediated transcription of miRNA precursor molecules containing a

5′-end cap structure and a 3′-end poly(A) tail. These long primary transcripts (pri-miRNAs) are cleaved by the Drosha–DGCR8 complex to produce 70 nt stem-loop structured precursors (pre-miRNAs), which are exported to the cytoplasm by exportin-5 and subsequently processed by Dicer [94]. Processing of the pre-miRNA by Dicer results in a mature miRNA guide strand that is loaded into the RNA-induced silencing complex (RISC) containing an Argonaute (Ago) protein and other RISC cofactors to form the microRNA-inducing silencing complex (miRISC) [94]. Mature miRISC targets specific mRNAs for translational repression or degradation [95]. Importantly, several components related to miRISC, such as miRNAs, mRNAs repressed by miRNAs, Ago proteins, DDX6, and Moloney leukemia virus 10 (MOV10), together with the mRNA degradation machinery, localize in PBs [96]. In fact, miRNA-mediated translational repression consistently correlates with mRNA accumulation in PBs [96].

One of the first works connecting HIV-1 with the cellular microRNA machinery was by Haase and colleagues, who described the identification of HIV-1 *trans*-activating response RNA-binding protein (TRBP) as a protein partner of human Dicer [97]. They showed that TRBP is required for optimal RNA silencing mediated by siRNAs and endogenous miRNAs, most probably by facilitating the cleavage of pre-miRNA [97] (Figure 2d). Then, Triboulet and colleagues demonstrated for the first time the physiological role of the miRNA-silencing machinery in controlling HIV-1 replication [98]. The authors showed that Type III RNAses Dicer and Drosha inhibited virus replication both in peripheral blood mononuclear cells from HIV-1-infected donors and in latently infected cells [98]. In turn, HIV-1 actively suppressed the expression of the polycistronic miRNA cluster miR-17/92, a miRNA cluster involved in genomic amplification in malignant lymphoma and lung cancer [99–101]. This specific suppression of the miR-17/92 cluster was found to be required for efficient viral replication and was dependent on p300/CBP-associated factor (PCAF), a histone acetyltransferase cofactor of TAT [98].

Subsequently, Nathans and colleagues reported that HIV-1 mRNA interacts with miRISC proteins and that disrupting PBs' structures resulted in enhanced viral production and infectivity [95]. The authors found that HIV-1 mRNAs are susceptible to targeting by the human miRNA miR-29a, which induces the association of viral mRNAs with miRISC. The authors also showed that miR-29a represses viral replication by inducing an accumulation of viral mRNA in PBs [95] (Figure 2d).

Another protein component of PBs shown to be important for miRNA-mediated repression that has been involved with the HIV replication cycle is MOV10. This protein belongs to the UPF1-like subfamily of DExD-box RNA helicases and has ATP-dependent 5′ to 3′ directional RNA helicase activity [102,103]. MOV10 was co-purified with APOBEC3G/A3G and shown to affect the assembly and maturation of miRISC [104]. In 2010, Burdick and colleagues reported that MOV10 overexpression resulted in reduced levels of both GAG protein and virus production [105]. The authors showed that MOV10 was efficiently incorporated into virions, reducing virus infectivity, in part, by interfering with reverse transcription [105]. In addition, MOV10 overexpression reduced the proteolytic processing of GAG by the viral protease and the authors showed that MOV10 specifically associates with HIV-1 unspliced mRNA [105]. Curiously, these authors showed that knockdown of MOV10 decreased virus production but showed little impact on virus infectivity, suggesting that basal levels of MOV10 are required for efficient viral replication [105]. Consistent with this last observation, Huang and colleagues demonstrated that MOV10 potently enhances nuclear export of viral mRNAs through the REV–RRE axis and subsequently increases the expression of GAG protein and other late products [106]. The authors also observed that MOV10 interacts with REV in an RNA-independent manner [106]. Given the discrepancies between both reports, further research is necessary to elucidate the role of MOV10 protein during the HIV-1 replication cycle.

As mentioned above, PBs are cytoplasmic foci associated with the mRNA decay machinery as they contain mRNA decapping enzymes (Dcp1/2) and the 5′-3′ exonuclease XRN1 [107,108], deadenylation factors (Ccr1, Caf1, Not1) [109], NMD-associated proteins (SMG5-6-7, UPF1) [107,110], and translational repressors (CPEB, eIF4E-T, DDX6) [111,112]. Co-localization of miRISC and target mRNAs in PBs suggests that they function in miRNA-mediated gene silencing by sequestering target mRNA for storage or decay [113–115]. Indeed, several PB components such as GW182 and DDX6 (RCK/p54) play important roles in miRNA-dependent translational repression [116].

Besides the interactions of HIV-1 with the miRISC machinery described above, several reports have shown that HIV-1 co-opts some PBs components to promote viral replication. It has been reported that depletion of Ago2 or DDX6 produces inhibition of HIV-1 replication, indicating a role of these PBs-associated proteins in the viral life cycle [117,118]. Indeed, Reed and colleagues demonstrated that the assembly intermediates (AIs), containing HIV-1 GAG, GAG–POL, and VIF [119], are formed by the recruitment of DDX6 and ATP-binding cassette protein E1 (ABCE1), thus providing evidence that HIV-1 utilizes these factors to catalyze the assembly of immature capsid intermediates [120] (Figure 2e). Interestingly, Abrahamyan and colleagues showed a dramatic decrease of PBs around HIV-1 unspliced mRNA-containing foci, suggesting a local dissolution of PBs close to assembly sites [89].

7. Interactions of HIV-1 and Components of RNA Granules Involved in mRNA Silencing

In response to environmental stress such as viral infection, eukaryotic cells reprogram their translational machinery to allow the selective expression of proteins required for cell viability in the face of changing conditions. mRNAs encoding constitutively expressed proteins are redirected from polysomes to RNA granules during stress conditions. Two of these RNA granules have been well characterized in yeast and mammalian cells, SGs, which correspond to translationally silent sites of RNA storage, and PBs, which are foci involved in mRNA degradation [121]. During stress, SG assembly signaling can be triggered by the phosphorylation of translation initiation factor eIF2α, which reduces the availability of the eIF2–GTP–tRNA^{Met} ternary complex necessary during translation initiation [122,123]. Interestingly, HIV-1 replication was shown to interfere with SG assembly in favor of the assembly of viral specific mRNP containing STAU1 [89]. More recently, Valiente-Echeverría and colleagues demonstrated that the HIV-1 GAG protein blocks SG assembly through an interaction between the N-terminal domain (NTD) of the capsid domain and the host eukaryotic elongation factor 2 (eEF2) [124] (Figure 2f). The authors also reported that GAG could mediate the disassembly of pre-existing SGs via an interaction with the SGs-dependency factor GTPase activating protein (GAP) SH3 domain-binding protein 1 (G3BP1) [124]. Interestingly, the mechanism by which HIV-1 interferes with SG assembly depends on the nature of the stressor. Indeed, the blockade of selenite-induced SGs was dependent on activation of eukaryotic translation initiation factor 4E binding protein 1 (4E-BP1) and the consequent inhibition of cap-binding by eIF4E [125] (Figure 2f). More recent data showed that G3BP1 binds the HIV-1 unspliced mRNA in the cytoplasm of macrophages to inhibit viral replication, supporting a role for G3BP1 and probably SGs as restriction factors that must be counteracted by HIV-1 [126].

8. Control of HIV-1 mRNA Abundance by Methylation of Viral Transcripts

Although hundreds of chemical modifications have been described in RNA, much less is known regarding the mechanisms and functions of these marks [127]. Methylation at the N^6 position of adenosines is the most abundant internal modification identified in mRNAs and lncRNAs in many eukaryotic species, including yeast and mammals [128]. Accumulating evidence suggests that m^{6}A regulates mRNA metabolism post-transcriptionally by altering the processing, nuclear export, translation, or stability of the modified mRNA [128].

It has been known for almost 40 years that, in addition to cellular mRNA, the RNAs of the influenza virus, adenovirus, Rous sarcoma virus (RSV), herpes simplex virus type 1 (HSV-1), and simian virus 40 (SV40) are m^{6}A-modified [129–133]. Although the precise impact of this modification on viral replication still remains unclear, recent studies revealed that the presence of the m^{6}A modification in the HIV-1 unspliced mRNA significantly affects gene expression and viral replication [134–136].

The post-transcriptional addition of m^{6}A to mRNAs occurs predominantly in the nucleus and is catalyzed by a heterotrimeric protein complex consisting of two methyltransferase-like enzymes, METTL3 and METTL14, and the cofactor Wilms' tumor 1-associated protein (WTAP), which together are recognized as the "writers" of m^6A [137–139]. Fat mass and obesity-associated protein (FTO) and α-ketoglutarate-dependent dioxygenase homolog 5 (ALKBH5) are two specific m^6A RNA demethylases ("erasers" of m^6A) responsible for adenosine demethylation and its associated regulation [140,141]. The main "readers" of m^6A on mRNAs are members of the so-called YTH domain-containing family. Amongst them, the cytoplasmic YTH proteins YTHDF1, YTHDF2, and YTHDF3, and the nuclear protein YTHDC1, have been shown to bind directly to m^6A-containing mRNAs [13,14,139,142]. Recent studies indicated that m^6A binding by YTHDF1 results in enhanced translational rates of its targets due to a specific interaction between YTHDF1 and eukaryotic initiation factor 3 (eIF3) [13]. In contrast, binding of m^6A by YTHDF2 results in both the localization of its mRNA targets in PBs and concomitant accelerated degradation [143]. m^6A-mediated mRNA degradation was shown to occur by the interaction between YTHDF2 and the CCR4–NOT deadenylase complex [128]. The precise function of YTHDF3 is still unclear [13,14,144]. Besides the mRNA degradation induced by binding of YTHDF2 to m^6A, there is also evidence suggesting that the presence of this chemical modification could indirectly destabilize some transcripts by preventing the binding of the mRNA stabilizing protein human antigen R (HuR) [145]. Moreover, m^6A regulates mRNA alternative splicing both directly through the recruitment of the m^6A reader YTHDC1 and indirectly by altering RNA structures close to the binding sites of the splicing factor heterogeneous nuclear ribonucleoprotein C (C1/C2) (hnRNPC) [142,146].

Recently, Lichinchi and colleagues reported that the HIV-1 unspliced mRNAs (and probably other viral transcripts) contain multiple m^6A modifications along their sequences [135]. Interestingly, the authors also observed that viral infection in a CD4+ T-cell line resulted in increased m^6A levels in cellular poly(A) RNA [135]. They also showed that methylation of two conserved adenosines within the stem loop II region of the RRE was important for binding of REV, resulting in enhanced rates of nuclear export of the methylated viral mRNA, thus revealing the importance of the m^6A modification for nuclear export of Rev-dependent viral mRNAs [135]. Moreover, Kennedy and colleagues found four clusters of m^6A modifications in the 3′-UTR region of the HIV-1 unspliced mRNA that enhanced viral gene expression by recruiting the three cytoplasmic m^6A "readers" proteins YTHDF1, 2, and 3 [134]. Notably, the authors observed that HIV-1 replication was dependent on the levels of YTHDF2 expression in infected T cells. As such, the virus presented enhanced GAG synthesis and viral particle production when YTHDF2 was overexpressed, while GAG protein and viral titers were reduced when the YTHDF2 gene was knocked out by the clustered regularly interspaced short palindromic repeats (CRISPR)/CRISPR-associated protein-9 nuclease (Cas9) system [134]. Contrary to the roles of YTHDF2 in mRNA degradation, the effects of YTHDF2 on HIV-1 replication were associated with enhanced viral mRNA abundance. Together, these data identified m^6A and the resultant recruitment of REV or YTHDF proteins as major positive regulators of HIV-1 mRNA abundance in the cytoplasm (Figure 2g) [134]. It is noteworthy that, similar to what has been reported for mRNA decay factors UPF1 and STAU1, these data suggest that HIV-1 uses the mRNA decay-associated m^6A reader YTHDF2 to promote viral replication.

More recently, Tirumuru and colleagues showed that proteins YTHDF1–3 recognize incoming m^6A-modified HIV-1 RNA and inhibit reverse transcription during the early steps of infection in both cell lines and primary CD4+ T-cells. Consistent with this observation, knockdown of YTHDF1–3 in cells resulted in enhanced reverse transcription products [136]. However, the same authors showed that silencing m^6A writers decreased HIV-1 GAG protein expression in virus-producer cells, while silencing m^6A erasers increased GAG expression. The authors concluded that m^6A plays a negative role during reverse transcription and a positive role later during viral gene expression [136].

Further research is necessary to elucidate the precise role of YTHDFs proteins, particularly YTHDF2, during the HIV-1 replicative cycle.

9. Conclusions and Future Perspectives

The HIV-1 unspliced mRNA plays critical roles during viral replication since it is used (i) as the precursor mRNA molecule undergoing alternative splicing in order to generate the remaining viral transcripts; (ii) as the mRNA template for GAG and GAG–POL synthesis; and (iii) as the genome packaged into newly assembled viral particles.

Interestingly, this 9 kb viral transcript possesses retained introns and AU-rich sequences—both signatures being incompatible with nuclear export and mRNA stability. In addition, the unspliced mRNA recruits different host factors including UPF1, STAU1/2, and the recently characterized m^6A reader protein YTHDF2, all of them associated with mRNA degradation. Despite all these constraints, HIV-1 has evolved mechanisms that ensure the presence and stability of the unspliced mRNA in the cytoplasm of host cells. The viral protein REV appears as a key factor, not only allowing the exit of the unspliced RNA from the nucleus by an alternative pathway and overcoming surveillance mechanisms but also interfering with *cis*-acting instability RNA elements and coordinating the recruitment of some of these mRNA decay factors that instead play positive roles in viral gene expression and virus production. The relationship between HIV-1 mRNAs and the host mRNA decay machinery has historically been a very poorly explored field. Whether viral proteins or the infection per se interfere with NMD, SMD, mRNA decapping, or deadenylation has to our knowledge never been evaluated. Further studies on this unexplored topic will help us to better understand the RNA biology behind HIV-1 replication and will certainly contribute to the development of new and novel drugs aimed at counteracting viral production and avoiding viral resistance.

Acknowledgments: Research at RSR laboratory is funded by the Comisión Nacional de Investigación en Ciencia y Tecnología (CONICYT) through the FONDECYT Program (Grant No. 1160176), the International Cooperation Program (DRI USA2013-0005) and the Associative Research Program (Anillo Grant ACT-1408). DTA holds a FONDECYT Postdoctoral fellow (Grant No. 3160091) and FVE holds a FONDECYT Initiation into Research Grant (No. 11140502).

Author Contributions: All authors contributed to the discussion and writing of this manuscript and approved the final version.

Conflicts of Interest: The authors declare no conflict of interest.

References

1. Hogg, J.R. Viral evasion and manipulation of host RNA quality control pathways. *J. Virol.* **2016**, *90*, 7010–7018. [CrossRef] [PubMed]
2. Davidson, L.; Kerr, A.; West, S. Co-transcriptional degradation of aberrant pre-mRNA by Xrn2. *EMBO J.* **2012**, *31*, 2566–2578. [CrossRef] [PubMed]
3. Hilleren, P.; McCarthy, T.; Rosbash, M.; Parker, R.; Jensen, T.H. Quality control of mRNA 3'-end processing is linked to the nuclear exosome. *Nature* **2001**, *413*, 538–542. [CrossRef] [PubMed]
4. Bousquet-Antonelli, C.; Presutti, C.; Tollervey, D. Identification of a regulated pathway for nuclear pre-mRNA turnover. *Cell* **2000**, *102*, 765–775. [CrossRef]
5. Decker, C.J.; Parker, R. P-bodies and stress granules: Possible roles in the control of translation and mRNA degradation. *Cold Spring Harb. Perspect. Biol.* **2012**, *4*, a012286. [CrossRef] [PubMed]
6. Doma, M.K.; Parker, R. Endonucleolytic cleavage of eukaryotic mRNAs with stalls in translation elongation. *Nature* **2006**, *440*, 561–564. [CrossRef] [PubMed]
7. Harigaya, Y.; Parker, R. No-go decay: A quality control mechanism for RNA in translation. *Wiley Interdiscip. Rev. RNA* **2010**, *1*, 132–141. [CrossRef] [PubMed]
8. Frischmeyer, P.A.; van Hoof, A.; O'Donnell, K.; Guerrerio, A.L.; Parker, R.; Dietz, H.C. An mRNA surveillance mechanism that eliminates transcripts lacking termination codons. *Science* **2002**, *295*, 2258–2261. [CrossRef] [PubMed]
9. Chen, C.Y.; Shyu, A.B. Selective degradation of early-response-gene mRNAs: Functional analyses of sequence features of the AU-rich elements. *Mol. Cell. Biol.* **1994**, *14*, 8471–8482. [CrossRef] [PubMed]
10. Wilusz, C.J.; Wormington, M.; Peltz, S.W. The cap-to-tail guide to mRNA turnover. *Nat. Rev. Mol. Cell Biol.* **2001**, *2*, 237–246. [CrossRef] [PubMed]

11. Chen, C.Y.; Shyu, A.B. AU-rich elements: Characterization and importance in mRNA degradation. *Trends Biochem. Sci.* **1995**, *20*, 465–470. [CrossRef]

12. Chen, C.Y.A.; Shyu, A.B. Emerging themes in regulation of global mRNA turnover in *cis*. *Trends Biochem. Sci.* **2016**. [CrossRef] [PubMed]

13. Wang, X.; Zhao, B.S.; Roundtree, I.A.; Lu, Z.; Han, D.; Ma, H.; Weng, X.; Chen, K.; Shi, H.; He, C. N^6-methyladenosine modulates messenger RNA translation efficiency. *Cell* **2015**, *161*, 1388–1399. [CrossRef] [PubMed]

14. Wang, X.; Lu, Z.; Gomez, A.; Hon, G.C.; Yue, Y.; Han, D.; Fu, Y.; Parisien, M.; Dai, Q.; Jia, G.; et al. N^6-methyladenosine-dependent regulation of messenger RNA stability. *Nature* **2014**, *505*, 117–120. [CrossRef] [PubMed]

15. Poblete-Duran, N.; Prades-Perez, Y.; Vera-Otarola, J.; Soto-Rifo, R.; Valiente-Echeverria, F. Who regulates whom? An overview of RNA granules and viral infections. *Viruses* **2016**, *8*, 180. [CrossRef] [PubMed]

16. Cassan, E.; Arigon-Chifolleau, A.M.; Mesnard, J.M.; Gross, A.; Gascuel, O. Concomitant emergence of the antisense protein gene of HIV-1 and of the pandemic. *Proc. Natl. Acad. Sci. USA* **2016**, *113*, 11537–11542. [CrossRef] [PubMed]

17. Frankel, A.D.; Young, J.A. HIV-1: Fifteen proteins and an RNA. *Annu. Rev. Biochem.* **1998**, *67*, 1–25. [CrossRef] [PubMed]

18. Karn, J.; Stoltzfus, C.M. Transcriptional and posttranscriptional regulation of HIV-1 gene expression. *Cold Spring Harb. Perspect. Med.* **2012**, *2*, a006916. [CrossRef] [PubMed]

19. Kim, S.Y.; Byrn, R.; Groopman, J.; Baltimore, D. Temporal aspects of DNA and RNA synthesis during human immunodeficiency virus infection: Evidence for differential gene expression. *J. Virol.* **1989**, *63*, 3708–3713. [PubMed]

20. Pomerantz, R.J.; Trono, D.; Feinberg, M.B.; Baltimore, D. Cells nonproductively infected with HIV-1 exhibit an aberrant pattern of viral RNA expression: A molecular model for latency. *Cell* **1990**, *61*, 1271–1276. [CrossRef]

21. Purcell, D.F.; Martin, M.A. Alternative splicing of human immunodeficiency virus type 1 mRNA modulates viral protein expression, replication, and infectivity. *J. Virol.* **1993**, *67*, 6365–6378. [PubMed]

22. Rojas-Araya, B.; Ohlmann, T.; Soto-Rifo, R. Translational control of the HIV unspliced genomic RNA. *Viruses* **2015**, *7*, 4326–4351. [CrossRef] [PubMed]

23. Rajanala, K.; Nandicoori, V.K. Localization of nucleoporin Tpr to the nuclear pore complex is essential for Tpr mediated regulation of the export of unspliced RNA. *PLoS ONE* **2012**, *7*, e29921. [CrossRef] [PubMed]

24. Coyle, J.H.; Bor, Y.C.; Rekosh, D.; Hammarskjold, M.L. The Tpr protein regulates export of mRNAs with retained introns that traffic through the Nxf1 pathway. *RNA* **2011**, *17*, 1344–1356. [CrossRef] [PubMed]

25. Malim, M.H.; Hauber, J.; Le, S.Y.; Maizel, J.V.; Cullen, B.R. The HIV-1 Rev *trans*-activator acts through a structured target sequence to activate nuclear export of unspliced viral mRNA. *Nature* **1989**, *338*, 254–257. [CrossRef] [PubMed]

26. Daly, T.J.; Cook, K.S.; Gray, G.S.; Maione, T.E.; Rusche, J.R. Specific binding of HIV-1 recombinant Rev protein to the Rev-Responsive Element in vitro. *Nature* **1989**, *342*, 816–819. [CrossRef] [PubMed]

27. Fornerod, M.; Ohno, M.; Yoshida, M.; Mattaj, I.W. CRM1 is an export receptor for leucine-rich nuclear export signals. *Cell* **1997**, *90*, 1051–1060. [CrossRef]

28. Henderson, B.R.; Percipalle, P. Interactions between HIV Rev and nuclear import and export factors: The Rev nuclear localisation signal mediates specific binding to human importin-beta. *J. Mol. Biol.* **1997**, *274*, 693–707. [CrossRef] [PubMed]

29. Arnold, M.; Nath, A.; Hauber, J.; Kehlenbach, R.H. Multiple Importins function as nuclear transport receptors for the Rev protein of human immunodeficiency virus type 1. *J. Biol. Chem.* **2006**, *281*, 20883–20890. [CrossRef] [PubMed]

30. Fischer, U.; Huber, J.; Boelens, W.C.; Mattaj, I.W.; Luhrmann, R. The HIV-1 Rev activation domain is a nuclear export signal that accesses an export pathway used by specific cellular RNAs. *Cell* **1995**, *82*, 475–483. [CrossRef]

31. Pollard, V.W.; Malim, M.H. The HIV-1 Rev protein. *Annu. Rev. Microbiol.* **1998**, *52*, 491–532. [CrossRef] [PubMed]

32. Chang, D.D.; Sharp, P.A. Regulation by HIV Rev depends upon recognition of splice sites. *Cell* **1989**, *59*, 789–795. [CrossRef]

33. Mann, D.A.; Mikaelian, I.; Zemmel, R.W.; Green, S.M.; Lowe, A.D.; Kimura, T.; Singh, M.; Butler, P.J.; Gait, M.J.; Karn, J. A molecular rheostat. Co-operative Rev binding to stem I of the Rev-response element modulates human immunodeficiency virus type-1 late gene expression. *J. Mol. Biol.* **1994**, *241*, 193–207. [CrossRef] [PubMed]

34. Rausch, J.W.; Le Grice, S.F. HIV Rev assembly on the Rev response element (RRE): A structural perspective. *Viruses* **2015**, *7*, 3053–3075. [CrossRef] [PubMed]

35. Cullen, B.R. Nuclear mRNA export: Insights from virology. *Trends Biochem. Sci.* **2003**, *28*, 419–424. [CrossRef]

36. Suhasini, M.; Reddy, T.R. Cellular proteins and HIV-1 Rev function. *Curr. HIV Res.* **2009**, *7*, 91–100. [CrossRef] [PubMed]

37. Kula, A.; Gharu, L.; Marcello, A. HIV-1 pre-mRNA commitment to Rev mediated export through PSF and Matrin 3. *Virology* **2013**, *435*, 329–340. [CrossRef] [PubMed]

38. Jones, T.; Sheer, D.; Bevec, D.; Kappel, B.; Hauber, J.; Steinkasserer, A. The human HIV-1 Rev binding-protein hRIP/Rab (HRB) maps to chromosome 2q36. *Genomics* **1997**, *40*, 198–199. [CrossRef] [PubMed]

39. Neville, M.; Stutz, F.; Lee, L.; Davis, L.I.; Rosbash, M. The importin-beta family member Crm1p bridges the interaction between Rev and the nuclear pore complex during nuclear export. *Curr. Biol.* **1997**, *7*, 767–775. [CrossRef]

40. Yedavalli, V.S.; Neuveut, C.; Chi, Y.H.; Kleiman, L.; Jeang, K.T. Requirement of DDX3 DEAD box RNA helicase for HIV-1 Rev-RRE export function. *Cell* **2004**, *119*, 381–392. [CrossRef] [PubMed]

41. Fang, J.; Kubota, S.; Yang, B.; Zhou, N.; Zhang, H.; Godbout, R.; Pomerantz, R.J. A DEAD box protein facilitates HIV-1 replication as a cellular co-factor of Rev. *Virology* **2004**, *330*, 471–480. [CrossRef] [PubMed]

42. Li, J.; Tang, H.; Mullen, T.M.; Westberg, C.; Reddy, T.R.; Rose, D.W.; Wong-Staal, F. A role for RNA helicase A in post-transcriptional regulation of HIV type 1. *Proc. Natl. Acad. Sci. USA* **1999**, *96*, 709–714. [CrossRef] [PubMed]

43. Felber, B.K.; Hadzopoulou-Cladaras, M.; Cladaras, C.; Copeland, T.; Pavlakis, G.N. Rev protein of human immunodeficiency virus type 1 affects the stability and transport of the viral mRNA. *Proc. Natl. Acad. Sci. USA* **1989**, *86*, 1495–1499. [CrossRef] [PubMed]

44. Hadzopoulou-Cladaras, M.; Felber, B.K.; Cladaras, C.; Athanassopoulos, A.; Tse, A.; Pavlakis, G.N. The *rev* (*trs/art*) protein of human immunodeficiency virus type 1 affects viral mRNA and protein expression via a *cis*-acting sequence in the *env* region. *J. Virol.* **1989**, *63*, 1265–1274. [PubMed]

45. Maldarelli, F.; Martin, M.A.; Strebel, K. Identification of posttranscriptionally active inhibitory sequences in human immunodeficiency virus type 1 RNA: Novel level of gene regulation. *J. Virol.* **1991**, *65*, 5732–5743. [PubMed]

46. Schwartz, S.; Felber, B.K.; Pavlakis, G.N. Distinct RNA sequences in the gag region of human immunodeficiency virus type 1 decrease RNA stability and inhibit expression in the absence of Rev protein. *J. Virol.* **1992**, *66*, 150–159. [PubMed]

47. Afonina, E.; Neumann, M.; Pavlakis, G.N. Preferential binding of poly(A)-binding protein 1 to an inhibitory RNA element in the human immunodeficiency virus type 1 gag mRNA. *J. Biol. Chem.* **1997**, *272*, 2307–2311. [PubMed]

48. Nasioulas, G.; Zolotukhin, A.S.; Tabernero, C.; Solomin, L.; Cunningham, C.P.; Pavlakis, G.N.; Felber, B.K. Elements distinct from human immunodeficiency virus type 1 splice sites are responsible for the Rev dependence of *env* mRNA. *J. Virol.* **1994**, *68*, 2986–2993. [PubMed]

49. Cochrane, A.W.; Jones, K.S.; Beidas, S.; Dillon, P.J.; Skalka, A.M.; Rosen, C.A. Identification and characterization of intragenic sequences which repress human immunodeficiency virus structural gene expression. *J. Virol.* **1991**, *65*, 5305–5313. [PubMed]

50. Schwartz, S.; Campbell, M.; Nasioulas, G.; Harrison, J.; Felber, B.K.; Pavlakis, G.N. Mutational inactivation of an inhibitory sequence in human immunodeficiency virus type 1 results in Rev-independent *gag* expression. *J. Virol.* **1992**, *66*, 7176–7182. [PubMed]

51. Zolotukhin, A.S.; Michalowski, D.; Bear, J.; Smulevitch, S.V.; Traish, A.M.; Peng, R.; Patton, J.; Shatsky, I.N.; Felber, B.K. PSF acts through the human immunodeficiency virus type 1 mRNA instability elements to regulate virus expression. *Mol. Cell. Biol.* **2003**, *23*, 6618–6630. [CrossRef] [PubMed]

52. Schneider, R.; Campbell, M.; Nasioulas, G.; Felber, B.K.; Pavlakis, G.N. Inactivation of the human immunodeficiency virus type 1 inhibitory elements allows Rev-independent expression of Gag and Gag/protease and particle formation. *J. Virol.* **1997**, *71*, 4892–4903. [PubMed]

53. Hentze, M.W. Determinants and regulation of cytoplasmic mRNA stability in eukaryotic cells. *Biochim. Biophys. Acta* **1991**, *1090*, 281–292. [CrossRef]

54. Shyu, A.B.; Greenberg, M.E.; Belasco, J.G. The c-*fos* transcript is targeted for rapid decay by two distinct mRNA degradation pathways. *Genes Dev.* **1989**, *3*, 60–72. [CrossRef] [PubMed]

55. Solomin, L.; Felber, B.K.; Pavlakis, G.N. Different sites of interaction for Rev, Tev, and Rex proteins within the Rev-Responsive Element of human immunodeficiency virus type 1. *J. Virol.* **1990**, *64*, 6010–6017. [PubMed]

56. Shaw, G.; Kamen, R. A conserved AU sequence from the 3′ untranslated region of GM-CSF mRNA mediates selective mRNA degradation. *Cell* **1986**, *46*, 659–667. [CrossRef]

57. Black, A.C.; Luo, J.; Watanabe, C.; Chun, S.; Bakker, A.; Fraser, J.K.; Morgan, J.P.; Rosenblatt, J.D. Polypyrimidine tract-binding protein and heterogeneous nuclear ribonucleoprotein A1 bind to human T-cell leukemia virus type 2 RNA regulatory elements. *J. Virol.* **1995**, *69*, 6852–6858. [PubMed]

58. Black, A.C.; Luo, J.; Chun, S.; Bakker, A.; Fraser, J.K.; Rosenblatt, J.D. Specific binding of polypyrimidine tract binding protein and hnRNP a1 to HIV-1 CRS elements. *Virus Genes* **1996**, *12*, 275–285. [CrossRef] [PubMed]

59. Valiente-Echeverria, F.; Vallejos, M.; Monette, A.; Pino, K.; Letelier, A.; Huidobro-Toro, J.P.; Mouland, A.J.; Lopez-Lastra, M. A *cis*-acting element present within the *gag* open reading frame negatively impacts on the activity of the HIV-1 IRES. *PLoS ONE* **2013**, *8*, e56962. [CrossRef] [PubMed]

60. Maquat, L.E.; Tarn, W.Y.; Isken, O. The pioneer round of translation: Features and functions. *Cell* **2010**, *142*, 368–374. [CrossRef] [PubMed]

61. Huang, L.; Wilkinson, M.F. Regulation of nonsense-mediated mRNA decay. *Wiley Interdiscip. Rev. RNA* **2012**, *3*, 807–828. [CrossRef] [PubMed]

62. Hoshino, S. Mechanism of the initiation of mRNA decay: Role of eRF3 family G proteins. *Wiley Interdiscip. Rev. RNA* **2012**, *3*, 743–757. [CrossRef] [PubMed]

63. Schoenberg, D.R.; Maquat, L.E. Regulation of cytoplasmic mRNA decay. *Nat. Rev. Genet.* **2012**, *13*, 246–259. [CrossRef] [PubMed]

64. Ghosh, S.; Jacobson, A. RNA decay modulates gene expression and controls its fidelity. *Wiley Interdiscip. Rev. RNA* **2010**, *1*, 351–361. [CrossRef] [PubMed]

65. Fabian, M.R.; Sonenberg, N. The mechanics of miRNA-mediated gene silencing: A look under the hood of miRISC. *Nat. Struct. Mol. Biol.* **2012**, *19*, 586–593. [CrossRef] [PubMed]

66. Hwang, J.; Maquat, L.E. Nonsense-mediated mRNA decay (NMD) in animal embryogenesis: To die or not to die, that is the question. *Curr. Opin. Genet. Dev.* **2011**, *21*, 422–430. [CrossRef] [PubMed]

67. Peltz, S.W.; Brown, A.H.; Jacobson, A. mRNA destabilization triggered by premature translational termination depends on at least three *cis*-acting sequence elements and one *trans*-acting factor. *Genes Dev.* **1993**, *7*, 1737–1754. [CrossRef] [PubMed]

68. Schweingruber, C.; Rufener, S.C.; Zund, D.; Yamashita, A.; Muhlemann, O. Nonsense-mediated mRNA decay—Mechanisms of substrate mRNA recognition and degradation in mammalian cells. *Biochim. Biophys. Acta* **2013**, *1829*, 612–623. [CrossRef] [PubMed]

69. Lykke-Andersen, J.; Shu, M.D.; Steitz, J.A. Human UPF proteins target an mRNA for nonsense-mediated decay when bound downstream of a termination codon. *Cell* **2000**, *103*, 1121–1131. [CrossRef]

70. Serin, G.; Gersappe, A.; Black, J.D.; Aronoff, R.; Maquat, L.E. Identification and characterization of human orthologues to saccharomyces cerevisiae UPF2 protein and UPF3 protein (*Caenorhabditis elegans* SMG-4). *Mol. Cell. Biol.* **2001**, *21*, 209–223. [CrossRef] [PubMed]

71. He, F.; Jacobson, A. Nonsense-mediated mRNA decay: Degradation of defective transcripts is only part of the story. *Annu. Rev. Genet.* **2015**, *49*, 339–366. [CrossRef] [PubMed]

72. Karousis, E.D.; Nasif, S.; Muhlemann, O. Nonsense-mediated mRNA decay: Novel mechanistic insights and biological impact. *Wiley Interdiscip. Rev. RNA* **2016**, *7*, 661–682. [CrossRef] [PubMed]

73. Balistreri, G.; Horvath, P.; Schweingruber, C.; Zund, D.; McInerney, G.; Merits, A.; Muhlemann, O.; Azzalin, C.; Helenius, A. The host nonsense-mediated mRNA decay pathway restricts mammalian RNA virus replication. *Cell Host Microb.* **2014**, *16*, 403–411. [CrossRef] [PubMed]

74. Ajamian, L.; Abrahamyan, L.; Milev, M.; Ivanov, P.V.; Kulozik, A.E.; Gehring, N.H.; Mouland, A.J. Unexpected roles for UPF1 in HIV-1 RNA metabolism and translation. *RNA* **2008**, *14*, 914–927. [CrossRef] [PubMed]

75. Hogg, J.R.; Goff, S.P. Upf1 senses 3′ UTR length to potentiate mRNA decay. *Cell* **2010**, *143*, 379–389. [CrossRef] [PubMed]

76. Ajamian, L.; Abel, K.; Rao, S.; Vyboh, K.; Garcia-de-Gracia, F.; Soto-Rifo, R.; Kulozik, A.E.; Gehring, N.H.; Mouland, A.J. HIV-1 recruits UPF1 but excludes UPF2 to promote nucleocytoplasmic export of the genomic RNA. *Biomolecules* **2015**, *5*, 2808–2839. [CrossRef] [PubMed]

77. Serquina, A.K.; Das, S.R.; Popova, E.; Ojelabi, O.A.; Roy, C.K.; Gottlinger, H.G. UPF1 is crucial for the infectivity of human immunodeficiency virus type 1 progeny virions. *J. Virol.* **2013**, *87*, 8853–8861. [CrossRef] [PubMed]

78. Mocquet, V.; Durand, S.; Jalinot, P. How Retroviruses escape the Nonsense-Mediated mRNA Decay. *AIDS Res. Hum. Retrovir.* **2015**, *31*, 948–958. [CrossRef] [PubMed]

79. Cullen, B.R. Role and mechanism of action of the APOBECc3 family of antiretroviral resistance factors. *J. Virol.* **2006**, *80*, 1067–1076. [CrossRef] [PubMed]

80. Köhrmann, M.; Luo, M.; Kaether, C.; DesGroseillers, L.; Dotti, C.G.; Kiebler, M.A. Microtubule-dependent recruitment of Staufen-green fluorescent protein into large RNA-containing granules and subsequent dendritic transport in living hippocampal neurons. *Mol. Biol. Cell* **1999**, *10*, 2945–2953. [CrossRef] [PubMed]

81. Kiebler, M.A.; Hemraj, I.; Verkade, P.; Köhrmann, M.; Fortes, P.; Marión, R.M.; Ortín, J.; Dotti, C.G. The mammalian staufen protein localizes to the somatodendritic domain of cultured hippocampal neurons: Implications for its involvement in mRNA transport. *J. Neurosci.* **1999**, *19*, 288–297. [PubMed]

82. Ricci, E.P.; Kucukural, A.; Cenik, C.; Mercier, B.C.; Singh, G.; Heyer, E.E.; Ashar-Patel, A.; Peng, L.; Moore, M.J. Staufen1 senses overall transcript secondary structure to regulate translation. *Nat. Struct. Mol. Biol.* **2014**, *21*, 26–35. [CrossRef] [PubMed]

83. Sugimoto, Y.; Vigilante, A.; Darbo, E.; Zirra, A.; Militti, C.; D'Ambrogio, A.; Luscombe, N.M.; Ule, J. hiCLIP reveals the in vivo atlas of mRNA secondary structures recognized by Staufen 1. *Nature* **2015**, *519*, 491–494. [CrossRef] [PubMed]

84. Park, E.; Maquat, L.E. Staufen-mediated mRNA decay. *Wiley Interdiscip. Rev. RNA* **2013**, *4*, 423–435. [CrossRef] [PubMed]

85. Dugre-Brisson, S.; Elvira, G.; Boulay, K.; Chatel-Chaix, L.; Mouland, A.J.; DesGroseillers, L. Interaction of Staufen1 with the 5′ end of mRNA facilitates translation of these RNAs. *Nucleic Acids Res.* **2005**, *33*, 4797–4812. [CrossRef] [PubMed]

86. Chatel-Chaix, L.; Clement, J.F.; Martel, C.; Beriault, V.; Gatignol, A.; DesGroseillers, L.; Mouland, A.J. Identification of Staufen in the human immunodeficiency virus type 1 Gag ribonucleoprotein complex and a role in generating infectious viral particles. *Mol. Cell. Biol.* **2004**, *24*, 2637–2648. [CrossRef] [PubMed]

87. Chatel-Chaix, L.; Abrahamyan, L.; Frechina, C.; Mouland, A.J.; DesGroseillers, L. The host protein Staufen1 participates in human immunodeficiency virus type 1 assembly in live cells by influencing pr55Gag multimerization. *J. Virol.* **2007**, *81*, 6216–6230. [CrossRef] [PubMed]

88. Chatel-Chaix, L.; Boulay, K.; Mouland, A.J.; Desgroseillers, L. The host protein Staufen1 interacts with the pr55Gag zinc fingers and regulates HIV-1 assembly via its N-terminus. *Retrovirology* **2008**, *5*, 41. [CrossRef] [PubMed]

89. Abrahamyan, L.G.; Chatel-Chaix, L.; Ajamian, L.; Milev, M.P.; Monette, A.; Clement, J.F.; Song, R.; Lehmann, M.; DesGroseillers, L.; Laughrea, M.; et al. Novel Staufen1 ribonucleoproteins prevent formation of stress granules but favour encapsidation of HIV-1 genomic RNA. *J. Cell Sci.* **2010**, *123*, 369–383. [CrossRef] [PubMed]

90. Banerjee, A.; Benjamin, R.; Balakrishnan, K.; Ghosh, P.; Banerjee, S. Human protein Staufen-2 promotes HIV-1 proliferation by positively regulating RNA export activity of viral protein Rev. *Retrovirology* **2014**, *11*, 18. [CrossRef] [PubMed]

91. Mu, X.; Fu, Y.; Zhu, Y.; Wang, X.; Xuan, Y.; Shang, H.; Goff, S.P.; Gao, G. HIV-1 exploits the host factor RuvB-like 2 to balance viral protein expression. *Cell Host Microb.* **2015**, *18*, 233–242. [CrossRef] [PubMed]

92. Voinnet, O. RNA silencing: Small RNAs as ubiquitous regulators of gene expression. *Curr. Opin. Plant Biol.* **2002**, *5*, 444–451. [CrossRef]

93. Saumet, A.; Lecellier, C.H. Anti-viral RNA silencing: Do we look like plants? *Retrovirology* **2006**, *3*, 3. [CrossRef] [PubMed]

94. Kim, V.N. MicroRNA biogenesis: Coordinated cropping and dicing. *Nat. Rev. Mol. Cell Biol.* **2005**, *6*, 376–385. [CrossRef] [PubMed]

95. Nathans, R.; Chu, C.Y.; Serquina, A.K.; Lu, C.C.; Cao, H.; Rana, T.M. Cellular microRNA and P bodies modulate host-HIV-1 interactions. *Mol. Cell* **2009**, *34*, 696–709. [CrossRef] [PubMed]

96.   Pillai, R.S.; Bhattacharyya, S.N.; Filipowicz, W. Repression of protein synthesis by miRNAs: How many mechanisms? *Trends Cell Biol.* **2007**, *17*, 118–126. [CrossRef] [PubMed]

97.   Haase, A.D.; Jaskiewicz, L.; Zhang, H.; Laine, S.; Sack, R.; Gatignol, A.; Filipowicz, W. TRBP, a regulator of cellular PKR and HIV-1 virus expression, interacts with Dicer and functions in RNA silencing. *EMBO Rep.* **2005**, *6*, 961–967. [CrossRef] [PubMed]

98.   Triboulet, R.; Mari, B.; Lin, Y.L.; Chable-Bessia, C.; Bennasser, Y.; Lebrigand, K.; Cardinaud, B.; Maurin, T.; Barbry, P.; Baillat, V.; et al. Suppression of microRNA-silencing pathway by HIV-1 during virus replication. *Science* **2007**, *315*, 1579–1582. [CrossRef] [PubMed]

99.   Hayashita, Y.; Osada, H.; Tatematsu, Y.; Yamada, H.; Yanagisawa, K.; Tomida, S.; Yatabe, Y.; Kawahara, K.; Sekido, Y.; Takahashi, T. A polycistronic microRNA cluster, miR-17–92, is overexpressed in human lung cancers and enhances cell proliferation. *Cancer Res.* **2005**, *65*, 9628–9632. [CrossRef] [PubMed]

100. Tagawa, H.; Seto, M. A microRNA cluster as a target of genomic amplification in malignant lymphoma. *Leukemia* **2005**, *19*, 2013–2016. [CrossRef] [PubMed]

101. He, L.; Thomson, J.M.; Hemann, M.T.; Hernando-Monge, E.; Mu, D.; Goodson, S.; Powers, S.; Cordon-Cardo, C.; Lowe, S.W.; Hannon, G.J.; et al. A microRNA polycistron as a potential human oncogene. *Nature* **2005**, *435*, 828–833. [CrossRef] [PubMed]

102. Gregersen, L.H.; Schueler, M.; Munschauer, M.; Mastrobuoni, G.; Chen, W.; Kempa, S.; Dieterich, C.; Landthaler, M. MOV10 is a 5′ to 3′ RNA helicase contributing to UPF1 mRNA target degradation by translocation along 3′ UTRs. *Mol. Cell* **2014**, *54*, 573–585. [CrossRef] [PubMed]

103. Abudu, A.; Wang, X.; Dang, Y.; Zhou, T.; Xiang, S.H.; Zheng, Y.H. Identification of molecular determinants from moloney leukemia virus 10 homolog (MOV10) protein for virion packaging and anti-HIV-1 activity. *J. Biol. Chem.* **2012**, *287*, 1220–1228. [CrossRef] [PubMed]

104. Gallois-Montbrun, S.; Kramer, B.; Swanson, C.M.; Byers, H.; Lynham, S.; Ward, M.; Malim, M.H. Antiviral protein APOBEC3G localizes to ribonucleoprotein complexes found in P bodies and stress granules. *J. Virol.* **2007**, *81*, 2165–2178. [CrossRef] [PubMed]

105. Burdick, R.; Smith, J.L.; Chaipan, C.; Friew, Y.; Chen, J.; Venkatachari, N.J.; Delviks-Frankenberry, K.A.; Hu, W.S.; Pathak, V.K. P body-associated protein MOV10 inhibits HIV-1 replication at multiple stages. *J. Virol.* **2010**, *84*, 10241–10253. [CrossRef] [PubMed]

106. Huang, F.; Zhang, J.; Zhang, Y.; Geng, G.; Liang, J.; Li, Y.; Chen, J.; Liu, C.; Zhang, H. RNA helicase MOV10 functions as a co-factor of HIV-1 Rev to facilitate Rev/RRE-dependent nuclear export of viral mRNAs. *Virology* **2015**, *486*, 15–26. [CrossRef] [PubMed]

107. Ingelfinger, D.; Arndt-Jovin, D.J.; Luhrmann, R.; Achsel, T. The human LSm1–7 proteins colocalize with the mRNA-degrading enzymes DCP1/2 and Xrnl in distinct cytoplasmic foci. *RNA* **2002**, *8*, 1489–1501. [PubMed]

108. Yu, J.H.; Yang, W.H.; Gulick, T.; Bloch, K.D.; Bloch, D.B. Ge-1 is a central component of the mammalian cytoplasmic mRNA processing body. *RNA* **2005**, *11*, 1795–1802. [CrossRef] [PubMed]

109. Sheth, U.; Parker, R. Targeting of aberrant mRNAs to cytoplasmic processing bodies. *Cell* **2006**, *125*, 1095–1109. [CrossRef] [PubMed]

110. Durand, S.; Cougot, N.; Mahuteau-Betzer, F.; Nguyen, C.H.; Grierson, D.S.; Bertrand, E.; Tazi, J.; Lejeune, F. Inhibition of nonsense-mediated mRNA decay (NMD) by a new chemical molecule reveals the dynamic of NMD factors in P-bodies. *J. Cell Biol.* **2007**, *178*, 1145–1160. [CrossRef] [PubMed]

111. Wilczynska, A.; Aigueperse, C.; Kress, M.; Dautry, F.; Weil, D. The translational regulator CPEB1 provides a link between DCP1 bodies and stress granules. *J. Cell Sci.* **2005**, *118*, 981–992. [CrossRef] [PubMed]

112. Andrei, M.A.; Ingelfinger, D.; Heintzmann, R.; Achsel, T.; Rivera-Pomar, R.; Luhrmann, R. A role for eIF4e and eIF4e-transporter in targeting mRNPs to mammalian processing bodies. *RNA* **2005**, *11*, 717–727. [CrossRef] [PubMed]

113. Liu, J.; Rivas, F.V.; Wohlschlegel, J.; Yates, J.R., 3rd; Parker, R.; Hannon, G.J. A role for the P-body component GW182 in microRNA function. *Nat. Cell Biol.* **2005**, *7*, 1261–1266. [CrossRef] [PubMed]

114. Pillai, R.S.; Bhattacharyya, S.N.; Artus, C.G.; Zoller, T.; Cougot, N.; Basyuk, E.; Bertrand, E.; Filipowicz, W. Inhibition of translational initiation by Let-7 microRNA in human cells. *Science* **2005**, *309*, 1573–1576. [CrossRef] [PubMed]

115. Sen, G.L.; Blau, H.M. Argonaute 2/RISC resides in sites of mammalian mRNA decay known as cytoplasmic bodies. *Nat. Cell Biol.* **2005**, *7*, 633–636. [CrossRef] [PubMed]

116. Eulalio, A.; Behm-Ansmant, I.; Izaurralde, E. P bodies: At the crossroads of post-transcriptional pathways. *Nat. Rev. Mol. Cell Biol.* **2007**, *8*, 9–22. [CrossRef] [PubMed]

117. Bouttier, M.; Saumet, A.; Peter, M.; Courgnaud, V.; Schmidt, U.; Cazevieille, C.; Bertrand, E.; Lecellier, C.H. Retroviral GAG proteins recruit AGO2 on viral RNAs without affecting RNA accumulation and translation. *Nucleic Acids Res.* **2012**, *40*, 775–786. [CrossRef] [PubMed]

118. Reed, J.C.; Molter, B.; Geary, C.D.; McNevin, J.; McElrath, J.; Giri, S.; Klein, K.C.; Lingappa, J.R. HIV-1 Gag co-opts a cellular complex containing DDX6, a helicase that facilitates capsid assembly. *J. Cell Biol.* **2012**, *198*, 439–456. [CrossRef] [PubMed]

119. Zimmerman, C.; Klein, K.C.; Kiser, P.K.; Singh, A.R.; Firestein, B.L.; Riba, S.C.; Lingappa, J.R. Identification of a host protein essential for assembly of immature HIV-1 capsids. *Nature* **2002**, *415*, 88–92. [CrossRef] [PubMed]

120. Lingappa, J.R.; Dooher, J.E.; Newman, M.A.; Kiser, P.K.; Klein, K.C. Basic residues in the nucleocapsid domain of Gag are required for interaction of HIV-1 Gag with ABCE1 (HP68), a cellular protein important for HIV-1 capsid assembly. *J. Biol. Chem.* **2006**, *281*, 3773–3784. [CrossRef] [PubMed]

121. Kedersha, N.; Stoecklin, G.; Ayodele, M.; Yacono, P.; Lykke-Andersen, J.; Fritzler, M.J.; Scheuner, D.; Kaufman, R.J.; Golan, D.E.; Anderson, P. Stress granules and processing bodies are dynamically linked sites of mRNP remodeling. *J. Cell Biol.* **2005**, *169*, 871–884. [CrossRef] [PubMed]

122. Kedersha, N.L.; Gupta, M.; Li, W.; Miller, I.; Anderson, P. RNA-binding proteins TIA-1 and TIAR link the phosphorylation of eIF-2 alpha to the assembly of mammalian stress granules. *J. Cell Biol.* **1999**, *147*, 1431–1442. [CrossRef] [PubMed]

123. Kedersha, N.; Cho, M.R.; Li, W.; Yacono, P.W.; Chen, S.; Gilks, N.; Golan, D.E.; Anderson, P. Dynamic shuttling of TIA-1 accompanies the recruitment of mRNA to mammalian stress granules. *J. Cell Biol.* **2000**, *151*, 1257–1268. [CrossRef] [PubMed]

124. Valiente-Echeverria, F.; Melnychuk, L.; Vyboh, K.; Ajamian, L.; Gallouzi, I.E.; Bernard, N.; Mouland, A.J. eEF2 and Ras-GAP SH3 domain-binding protein (G3BP1) modulate stress granule assembly during HIV-1 infection. *Nat. Commun.* **2014**, *5*, 4819. [CrossRef] [PubMed]

125. Cinti, A.; Le Sage, V.; Ghanem, M.; Mouland, A.J. HIV-1 Gag blocks selenite-induced stress granule assembly by altering the mRNA cap-binding complex. *mBio* **2016**, *7*, e00329. [CrossRef] [PubMed]

126. Cobos Jimenez, V.; Martinez, F.O.; Booiman, T.; van Dort, K.A.; van de Klundert, M.A.; Gordon, S.; Geijtenbeek, T.B.; Kootstra, N.A. G3BP1 restricts HIV-1 replication in macrophages and T-cells by sequestering viral RNA. *Virology* **2015**, *486*, 94–104. [CrossRef] [PubMed]

127. Cantara, W.A.; Crain, P.F.; Rozenski, J.; McCloskey, J.A.; Harris, K.A.; Zhang, X.; Vendeix, F.A.; Fabris, D.; Agris, P.F. The RNA modification database, RNAMDB: 2011 Update. *Nucleic Acids Res.* **2011**, *39*, D195–D201. [CrossRef] [PubMed]

128. Du, H.; Zhao, Y.; He, J.; Zhang, Y.; Xi, H.; Liu, M.; Ma, J.; Wu, L. YTHDF2 destabilizes m⁶A-containing RNA through direct recruitment of the CCR4-NOT deadenylase complex. *Nat. Commun.* **2016**, *7*, 12626. [CrossRef] [PubMed]

129. Beemon, K.; Keith, J. Localization of N^6-methyladenosine in the rous sarcoma virus genome. *J. Mol. Biol.* **1977**, *113*, 165–179. [CrossRef]

130. Canaani, D.; Kahana, C.; Lavi, S.; Groner, Y. Identification and mapping of N^6-methyladenosine containing sequences in simian virus 40 RNA. *Nucleic Acids Res.* **1979**, *6*, 2879–2899. [CrossRef] [PubMed]

131. Hashimoto, S.I.; Green, M. Multiple methylated cap sequences in adenovirus type 2 early mRNA. *J. Virol.* **1976**, *20*, 425–435. [PubMed]

132. Krug, R.M.; Morgan, M.A.; Shatkin, A.J. Influenza viral mRNA contains internal N^6-methyladenosine and 5′-terminal 7-methylguanosine in cap structures. *J. Virol.* **1976**, *20*, 45–53. [PubMed]

133. Moss, B.; Gershowitz, A.; Stringer, J.R.; Holland, L.E.; Wagner, E.K. 5′-terminal and internal methylated nucleosides in herpes simplex virus type 1 mRNA. *J. Virol.* **1977**, *23*, 234–239. [PubMed]

134. Kennedy, E.M.; Bogerd, H.P.; Kornepati, A.V.; Kang, D.; Ghoshal, D.; Marshall, J.B.; Poling, B.C.; Tsai, K.; Gokhale, N.S.; Horner, S.M.; et al. Post-transcriptional m⁶A editing of HIV-1 mRNAs enhances viral gene expression. *Cell Host Microb.* **2016**, *19*, 675–685. [CrossRef] [PubMed]

135. Lichinchi, G.; Gao, S.; Saletore, Y.; Gonzalez, G.M.; Bansal, V.; Wang, Y.; Mason, C.E.; Rana, T.M. Dynamics of the human and viral m⁶A RNA methylomes during HIV-1 infection of t cells. *Nat. Microbiol.* **2016**, *1*, 16011. [CrossRef] [PubMed]

136. Tirumuru, N.; Zhao, B.S.; Lu, W.; Lu, Z.; He, C.; Wu, L. N^6-methyladenosine of HIV-1 RNA regulates viral infection and HIV-1 Gag protein expression. *eLife* **2016**, *5*, e15528. [CrossRef] [PubMed]

137. Liu, J.; Yue, Y.; Han, D.; Wang, X.; Fu, Y.; Zhang, L.; Jia, G.; Yu, M.; Lu, Z.; Deng, X.; et al. A METTL3-METTL14 complex mediates mammalian nuclear RNA N^6-adenosine methylation. *Nat. Chem. Biol.* **2014**, *10*, 93–95. [CrossRef] [PubMed]

138. Meyer, K.D.; Saletore, Y.; Zumbo, P.; Elemento, O.; Mason, C.E.; Jaffrey, S.R. Comprehensive analysis of mRNA methylation reveals enrichment in 3′ UTRs and near stop codons. *Cell* **2012**, *149*, 1635–1646. [CrossRef] [PubMed]

139. Yue, Y.; Liu, J.; He, C. RNA N^6-methyladenosine methylation in post-transcriptional gene expression regulation. *Genes Dev.* **2015**, *29*, 1343–1355. [CrossRef] [PubMed]

140. Jia, G.; Fu, Y.; Zhao, X.; Dai, Q.; Zheng, G.; Yang, Y.; Yi, C.; Lindahl, T.; Pan, T.; Yang, Y.G.; et al. N^6-methyladenosine in nuclear RNA is a major substrate of the obesity-associated FTO. *Nat. Chem. Biol.* **2011**, *7*, 885–887. [CrossRef] [PubMed]

141. Zheng, G.; Dahl, J.A.; Niu, Y.; Fedorcsak, P.; Huang, C.M.; Li, C.J.; Vagbo, C.B.; Shi, Y.; Wang, W.L.; Song, S.H.; et al. ALKBH5 is a mammalian RNA demethylase that impacts RNA metabolism and mouse fertility. *Mol. Cell* **2013**, *49*, 18–29. [CrossRef] [PubMed]

142. Xiao, W.; Adhikari, S.; Dahal, U.; Chen, Y.S.; Hao, Y.J.; Sun, B.F.; Sun, H.Y.; Li, A.; Ping, X.L.; Lai, W.Y.; et al. Nuclear m^{6}A reader YTHDC1 regulates mRNA splicing. *Mol. Cell* **2016**, *61*, 507–519. [CrossRef] [PubMed]

143. Li, F.; Zhao, D.; Wu, J.; Shi, Y. Structure of the YTH domain of human YTHDF2 in complex with an m^{6}A mononucleotide reveals an aromatic cage for m^{6}A recognition. *Cell Res.* **2014**, *24*, 1490–1492. [CrossRef] [PubMed]

144. Wang, X.; He, C. Reading RNA methylation codes through methyl-specific binding proteins. *RNA Biol.* **2014**, *11*, 669–672. [CrossRef] [PubMed]

145. Wang, Y.; Li, Y.; Toth, J.I.; Petroski, M.D.; Zhang, Z.; Zhao, J.C. N^6-methyladenosine modification destabilizes developmental regulators in embryonic stem cells. *Nat. Cell Biol.* **2014**, *16*, 191–198. [CrossRef] [PubMed]

146. Liu, N.; Dai, Q.; Zheng, G.; He, C.; Parisien, M.; Pan, T. N^6-methyladenosine-dependent RNA structural switches regulate RNA-protein interactions. *Nature* **2015**, *518*, 560–564. [CrossRef] [PubMed]

viruses

MDPI

Review

Gene Regulation and Quality Control in Murine Polyomavirus Infection

Gordon G. Carmichael

Department of Genetics and Genome Sciences, UCONN Health, Farmington, CT 06030, USA; carmichael@uchc.edu; Tel.: +1-860-679-2259

Academic Editors: J. Robert Hogg and Karen L. Beemon
Received: 19 September 2016; Accepted: 11 October 2016; Published: 17 October 2016

Abstract: Murine polyomavirus (MPyV) infects mouse cells and is highly oncogenic in immunocompromised hosts and in other rodents. Its genome is a small, circular DNA molecule of just over 5000 base pairs and it encodes only seven polypeptides. While seemingly simply organized, this virus has adopted an unusual genome structure and some unusual uses of cellular quality control pathways that, together, allow an amazingly complex and varied pattern of gene regulation. In this review we discuss how MPyV leverages these various pathways to control its life cycle.

Keywords: quality control; transcription; RNA decay; RNA editing; nuclear retention

1. The Virus

Murine polyomavirus (MPyV) is highly oncogenic in rodents and has a small circular double-stranded DNA (dsDNA) genome of about 5300 base pairs. The genome is divided into "early" and "late" regions, which are expressed and regulated differently as infection proceeds (Figure 1) [1–4]. The early and late transcription units extend in opposite directions around the circular genome from start sites near the bidirectional origin of DNA replication [2,5]. Primary RNA products from the early transcription unit are alternatively spliced to yield four early mRNAs which encode the large T antigen (100 kDa), the middle T antigen (56 kDa), the small T antigen (22 kDa) and the tiny T antigen (10 kDa) [6]. Large T binds to sequences in or near the DNA replication origin region [7–10] and is involved in the initiation of DNA replication, indirectly in the autoregulation of early-strand RNA levels [11–13], and indirectly in the activation of high levels of expression from the late promoter [13,14]. The other early proteins are dispensable for lytic infection, but are important for cell transformation and tumorigenesis [15]. Late primary transcripts accumulate after the onset of DNA replication and are also spliced in alternative ways to give mRNAs which code for the three virion structural proteins VP1, VP2 and VP3.

While seemingly simply organized, MPyV has adopted an unusual genome structure that provides a platform for the participation of a number of cellular gene regulatory and quality control mechanisms. First, the intergenic region is complex and crowded and serves multiple functions during infection. Consisting of only several hundred nucleotides, this region contains the origin of bidirectional DNA replication, the early promoter and the late promoter. Each of these is impacted by distinct molecular machinery, competing for overlapping sequence elements. Activation of the replication origin requires the recruitment of the cellular DNA replication machinery by large T antigen, which recognizes a number of sites in this region. The early promoter is a typical RNA polymerase II promoter, including a TATA box to specify early transcription start sites and an upstream enhancer region. The late promoter is TATA-less and specifies transcripts with a multitude of 5'-ends spanning more than 100 nucleotides. Second, the distal ends of the early and late regions are tightly connected (Figure 2) and, as we will see below, this organization plays a major role in the regulation of the viral life cycle. The ends of the coding regions for large T antigen and VP1 are very close to one another, separated from each other by only 23 base pairs. Also, the polyadenylation signals for early-strand and late-strand primary transcripts actually overlap one another. This leads to overlapping 3′-ends of early and late mRNAs and pre-mRNAs, with the amount of overlap being 45 base pairs or greater. As we shall discuss below, transcript overlap appears to be essential for the viral life cycle, since viruses that are constructed to eliminate this overlap fail to enter the late phase of infection [16]. Third, the splicing signals for late mRNAs are arranged in a manner rarely seen in eukaryotic transcripts. In most pre-mRNA molecules, the first splice site encountered is a donor, 5′-splice site. This allows the splicing of the first exon to the second exon. In MPyV late transcripts, the cap-proximal splice site is actually an acceptor, 3′-splice site. This almost unique arrangement turns out to be critical for the viral life cycle.

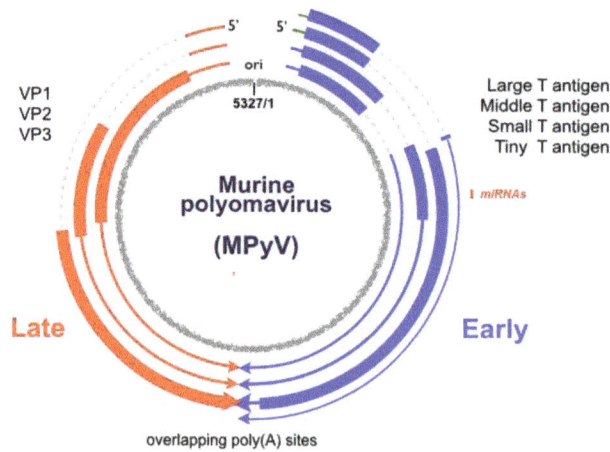

Figure 1. The murine polyomavirus (MPyV) genome. The genome shown is of strain NG59RA, which is 5327 base pairs in length. Early genes are in blue and late genes are in red. Transcripts are lines, with thicker regions denoting open reading frames and dotted lines introns. The replication origin and transcriptional control region is shown at the top of the genome. Late-strand transcripts can give rise to two miRNAs (small red line) that map to the early region and can influence viral and host gene expression.

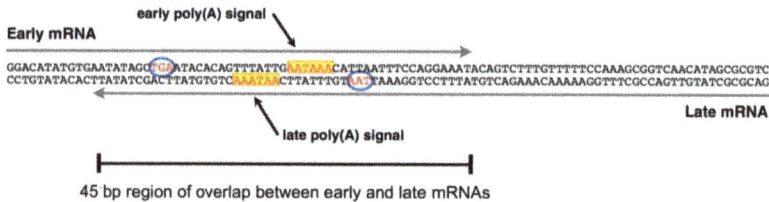

Figure 2. A crowded arrangement at the ends of the early and late genes. The sequence shown is of the 3′-region of the early and late transcription units, with the early coding strand on top and the stop codons for large T antigen and the virion structural protein VP1 circled. Note the overlap of the polyadenylation signals, including the canonical AATAAA elements (yellow box). Cleavage and polyadenylation occur downstream of these elements, leading to early and late mRNAs that have the potential to overlap for at least 45 base pairs (bp) at their 3′-ends. Transcript overlap is essential for the viral life cycle.

2. The Viral Early–Late Switch

Temporal regulation of MPyV gene expression during lytic infection of permissive mouse cells proceeds in a well-defined and tightly regulated manner [1,17,18]. Immediately after infection, RNA from the early transcription unit begins to accumulate; however, RNA from the late transcription unit fails to accumulate to a significant level. At 12 h after infection, the early–late RNA ratio is about 4 to 1 [1,18–20] and in the presence of DNA replication inhibitors, the ratio is 10 to 1 or even higher. At 12–15 h post-infection, viral DNA replication commences and late-strand RNA begins to accumulate rapidly and almost exponentially, while early-strand RNA and proteins accumulate much more slowly. In absolute terms, the amount of early-strand RNA in the cell is similar at 12 h and 24 h post infection. Thus, there is a dramatic change in the relative abundances of early-strand and late-strand RNAs; by 24 h post-infection, the early to late RNA ratio is as low as 1 to 50 [1,18–20]. This early–late "switch" depends absolutely on viral DNA replication. If replication is inhibited, early mRNAs continue to accumulate but late mRNAs fail to do so [11,12,19–22]. While it was thought in the field for a number of years that the early–late switch is the result of T antigen repression of the early promoter, coupled with a transactivation of the late promoter, this now seems to be incorrect. Rather, the switch appears to result from changes in transcription elongation and/or RNA stability [13,19,23–25]. Late RNA accumulation is regulated post-transcriptionally by what appears to be a novel RNA titration event (late gene expression from a non-replicating viral genome can be activated in *trans* by sufficient levels of late transcription from a replicating genome in the same cell) [13], while early RNA levels are regulated at least in part by antisense RNA and RNA editing (Figure 3) [13], as well as by virus-encoded miRNA [25] (see Figure 1).

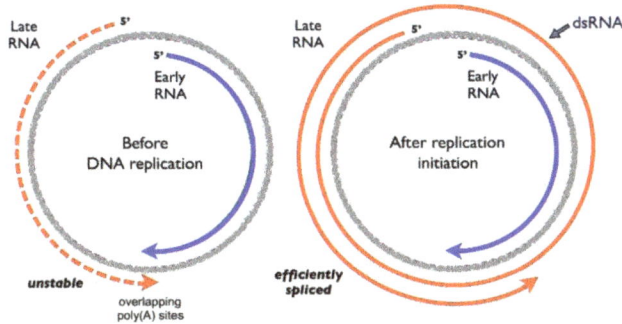

Figure 3. The early–late switch is associated with poly(A) signal readthrough and double-stranded RNA (dsRNA) formation. See text for details of the regulation. At early times and before viral DNA replication (**left**), transcription occurs from both the early and late promoters. Early-strand RNAs are spliced to produce mRNAs for the early proteins. Late-strand transcripts are efficiently terminated and polyadenylated, but are unstable and produce only small amounts of late mRNAs and proteins. After the onset of DNA replication (**right**), transcription termination and polyadenylation become less efficient, allowing multigenomic transcripts to be produced. Giant transcripts are efficiently spliced to generate stable late mRNAs, but sequences antisense to early-strand transcripts can downregulate early genes.

Even before the onset of viral DNA replication, however, the late-strand is actually being transcribed, but with little stable RNA accumulation. This latter phenomenon is associated with several important genomic features. At early times after infection, polyadenylation of late-strand transcripts is efficient. This generates RNAs that can be alternatively spliced to generate mRNAs for the virion structural proteins VP1 and VP3. VP2 mRNA from these transcripts is unspliced. Importantly, however, by a mechanism that remains unclear, these RNAs appear to be inefficiently exported from the nucleus to the cytoplasm and are degraded in the nucleus [26].

3. Late-Strand RNAs

How is late-strand gene expression enhanced at late times in infection? While at early times late-strand polyadenylation is efficient, this changes dramatically at late times. After DNA replication initiation, late-strand polyadenylation becomes inefficient, allowing RNA polymerase II to continue around and around the circular genome, generating giant multigenomic transcripts. Thus, the MPyV life cycle can be viewed as an interesting model of regulation of alternative polyadenylation, a phenomenon that has been studied in a variety of other systems [27,28]. Most late-strand primary transcripts are heterogeneous in size, and range from about 2.5 Kb to over 60 Kb in length [29–34]. Most are not polyadenylated [34]. Late-strand pre-mRNA molecules are processed into mature mRNAs using a highly unusual pathway that involves ordered splice site selection from precursors containing tandemly repeated introns and exons [35]. The great majority of late RNA sequences never leave the nucleus as they are removed during mRNA processing, and are subsequently degraded [30,36]. Some of these giant transcripts may also serve as precursors for the processing of viral miRNAs, one of which also downregulates the pro-apoptotic factor Smad2 [25,37].

The MPyV late region encodes 57-nucleotide non-coding exon near at the 5′-end of the transcription unit. At their 5′-ends, late messages contain multiple tandem repeats of this late leader sequence, which appears only once in the viral genome. Pre-mRNA molecules are processed by a pathway that includes the splicing of late leader exons to each other (Figure 4). Each class of late viral message (encoding virion structural proteins VP1, VP2 or VP3) consists of molecules with between 1 and 12 tandem leader units at the 5′-end [38], with most containing more than one. VP2 mRNA is the least abundant late message (about 5%) and contains no leader-to-body splice. Even in the absence of leader-to-body splicing, this message is nevertheless exported to the cytoplasm, although inefficiently [39]. Late-strand pre-mRNA processing is highly unusual, because it involves alternative selection between identical splice sites. Thus, in long pre-mRNAs, only the terminal coding body 3′-splice site is chosen, even though an upstream one exists in the precursors [35,38].

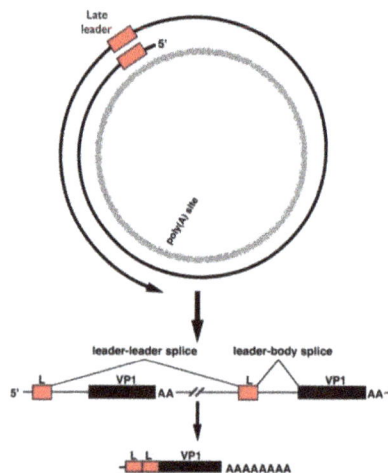

Figure 4. Late pre-mRNA splicing. Giant transcripts serve as precursors to late mRNAs. Processing of VP1 mRNA is shown. In multigenomic transcripts, leader (L) exons splice to one another, removing genome-length introns. Then, a leader-body splice can occur, coincident with polyadenylation. This results in mRNAs with tandem non-coding late leader exons at their 5′-ends.

While the splicing process is connected to mRNA accumulation, we hypothesize that tandem leaders may serve the additional purpose of facilitating translation initiation owing to the fact that leaders contain two regions with significant complementarity to the 3'-end of mouse 18S rRNA (Figure 5). Such regions could be coincidental and there may exist numerous other regions in cellular or viral RNAs. However, as most late MPyV mRNAs contain multiple tandem leaders in their 5'-untranslated regions (UTRs), this feature of multiple complementary sequences, preceding late AUG codons, could serve as a powerful way to recruit ribosomes and enhance the expression of virion structural proteins late in infection.

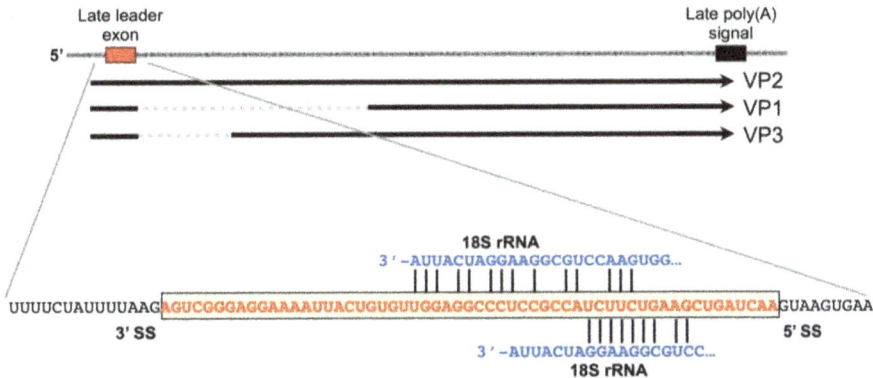

Figure 5. The late leader exon has regions of complementarity to 18S rRNA. While the biological consequence of this still remains unclear, there is striking complementarity to ribosomal RNA at two positions within the leader. We speculate that in late mRNAs containing tandem leaders in their 5'-untranslated regions (UTRs), these elements may serve to enhance the translation of late proteins.

4. Activation of Late RNA Accumulation

Late-strand gene expression may not be regulated primarily at the level of transcription initiation. Non-replicating genomes express only very low levels of late-strand transcripts. However, late genes from these non-replicating genomes are turned on if a replicating polyoma genome is introduced into the same cell [13]. Further, the presence of wild type genomes in mouse cells can lead to the activation of late genes in *trans* from a non-replicating genome in the same cells [13].

5. The Role of dsRNA Formation and A-To-I Editing in MPyV Gene Regulation

Due to the readthrough of early and late transcripts at late times, as well as the genomic overlap of the early and late polyadenylation signals, there is the possibility that if complementary sequences accumulate near one another in the nucleus, they might anneal to form double-stranded RNAs (dsRNAs). Also, since viruses that do not allow early-strand and late-strand overlap do not undergo productive infection and cannot enter a normal late phase [16], it is likely that dsRNA plays an important role. Nuclear dsRNAs can be promiscuously edited by dsRNA-specific adenosine deaminase (ADAR) enzymes, which deaminate adenosines to inosines [40]. Consistent with sense–antisense overlap of the MPyV early and late transcripts, viral RNAs also exhibit extensive and promiscuous editing [16,20,41,42]. During productive infection, there is a time-dependent increase in editing, with especially efficient editing observed around the overlapping polyadenylation sites [20]. No significant editing was detected before DNA replication or in the presence of a replication inhibitor [16,20]. Editing in the polyadenylation region has led to speculation that this editing serves as a trigger for the early–late switch. While editing is readily and abundantly observed, however, at this time we cannot conclude whether it is a cause or a consequence of viral gene regulation. The possibility exists that editing is a consequence of dsRNA formation while duplex RNA formation may in fact be the primary event that drives the early–late switch. This is because mouse cells lacking ADAR activity have been reported to maintain the ability to support productive MPyV infection [43].

6. Early-Strand RNAs

How is early-strand gene expression lowered at late times in infection? There appear to be multiple mechanisms for this. While the switch from early to late phase of infection has been reported by others to be regulated primarily at the level of transcription [44], this has been challenged by results which are, in fact, consistent with a change in the processing of late-strand transcripts [14,19]. Inefficient late-strand polyadenylation and transcription termination appear to regulate early-strand gene expression in an indirect manner. The long, multigenomic late-strand transcripts in the nucleus can form RNA–RNA duplexes with early-strand transcripts which are efficient substrates for RNA editing by ADARs. This leads to the deamination of up to 50% of the adenosines on each strand to inosines (which act biochemically and genetically like guanosines). These promiscuously edited RNAs are retained in the nucleus by a quality control system involving binding to the p54^{nrb}/NONO protein [45] and localization to nuclear bodies called paraspeckles [46,47], preventing them from being exported to the cytoplasm and being translated into mutant proteins [45]. As late primary transcripts accumulate to high levels in the nucleus, the opportunity for sense–antisense hybrids to form may increase, leading to ever greater inhibition by editing. While a similar phenomenon may also occur on the late strand (early-strand polyadenylation site readthrough, followed by dsRNA formation and A-to-I editing), the consequences in this case are minor because the editing would occur in genome-length introns of late pre-mRNAs rather than in late coding regions. In this manner, a cellular quality control system that prevents the nucleocytoplasmic export of dsRNAs and promiscuously edited RNAs plays an important role in reducing the availability of MPyV early-strand mRNAs for translation at late times in infection when the early gene products are no longer needed.

Yet another way in which early expression changes after the onset of viral DNA replication is at the level of transcription initiation. Through a mechanism that still remains unclear, as infection proceeds, the 5′ transcriptional start sites from the early promoter shift progressively further and further upstream from the canonical start site downstream of the early TATA element, which is used almost exclusively in the absence of DNA replication [20,22,48] (Figure 6). Thus, at late times, many early-strand mRNAs have 5′-UTRs that are hundreds of nucleotides longer than mRNAs at early times. This shift is dependent on DNA replication but not directly on the presence of large T antigen, because large T antigen is expressed in the presence of replication inhibitors, yet in this case the shift does not occur [20]. What are the consequences of shifting early-strand 5′-ends? They may alter RNA stability and therefore lead to reduced levels of mRNA. We hypothesize, however, that they serve as yet another mechanism to limit early gene expression at late times, by leading to inefficient ribosome scanning and translation initiation. We note that many of the early-strand mRNAs at late times contain AUG codons that are frameshifted relative to the normal AUG codon and therefore would be poor messages for T antigen expression. This replication-dependent switch has also been observed for several other viruses, suggesting a more common mechanism by which small DNA viruses might limit early gene expression late in infection. Altered early-strand start sites at late times have been reported both for simian virus 40 (SV40) [49] and for the John Cunningham (JC) virus [50].

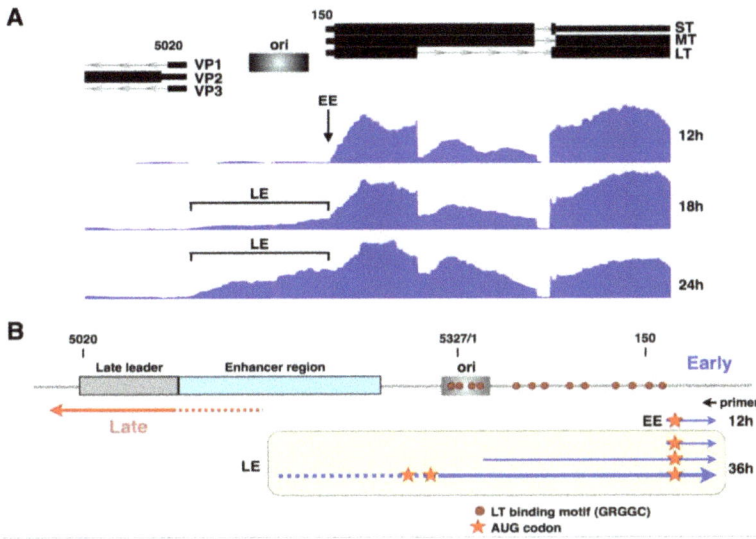

Figure 6. Early-strand transcription start sites shift after the onset of viral DNA replication. (**A**) An expanded view of the intergenic region is shown, along with genome browser tracks showing the alignment of early-strand RNAs at several times after infection, as reported by us recently [20]. These data were confirmed using the 5′-rapid amplification of cDNA ends (RACE) analysis [20]. Note the dramatic shift from 5′-ends mapping to a specific site at early times (EE) to many upstream sites at later times (LE); (**B**) The intergenic region is depicted, along with a general cartoon of early-strand RNAs at early times (EE) and early-strand RNAs at late times (LE). Positions of large T antigen binding are shown, along with the palindromic core replication origin, the enhancer region and the late transcription start site region. Red stars denote AUG codons that could direct translation initiation. Those in LE but not EE transcripts are frameshifted relative to the early coding region.

7. Conclusions

In conclusion, the mouse polyoma virus employs a variety of mechanisms to regulate the synthesis, processing, stability and translation of its RNAs in order to optimize the timing and efficiency of its life cycle. Taken together, the various modes of regulation adopted have given MPyV a powerful set of strategies to ensure efficient progression through its lytic life cycle. Some of these (such as shifting transcription start sites and promoter regulation) are shared by other viruses and systems, while others (such as the role of leader-to-leader splicing, polyadenylation site overlap and nuclear retention of dsRNAs and edited RNAs) are interesting and perhaps peculiar to this virus. Interestingly, while SV40 does not normally downregulate its early gene products using antisense RNA, it has been reported that in SV40-transformed human mesothelial cells, an integrated viral genome promotes polyadenylation site readthrough from the early region, thus generating antisense RNA that downregulates late gene expression [51]. Further, a recent study on the Merkel cell polyomavirus (MCPyV) life cycle presented transcriptomic data consistent with multigenomic transcripts, similar to those we see in MPyV infection [52].

Acknowledgments: I thank Kim Morris for helpful comments on the manuscript. This work was supported by grant CA045382 from the National Cancer Institute, NIH, USA.

Conflicts of Interest: The author declares no conflict of interest.

References

1. Kamen, R.; Lindstrom, D.M.; Shure, H.; Old, R.W. Virus-specific RNA in cells productively infected or transformed by polyoma virus. *Cold Spring Harb. Symp. Quant. Biol.* **1974**, *39*, 187–198. [CrossRef]
2. Griffin, B.; Fried, M. Amplification of a specific region of the polyoma virus genome. *Nature* **1975**, *256*, 175–179. [CrossRef] [PubMed]
3. Kamen, R.; Favaloro, J.; Parker, J. Topography of the three late mRNA's of polyoma virus which encode the virion proteins. *J. Virol.* **1980**, *33*, 637–651. [PubMed]
4. Kamen, R.; Favaloro, J.; Parker, J.; Treisman, R.; Lania, L.; Fried, M.; Mellor, A. A comparison of polyomavirus transcription in productively infected mouse cells and transformed rodent cell lines. *Cold Spring Harb. Symp. Quant. Biol.* **1980**, *44*, 63–75. [CrossRef] [PubMed]
5. Crawford, L.; Robbins, A.; Nicklin, P. Location of the origin and terminus of replication in polyoma virus DNA. *J. Gen. Virol.* **1974**, *25*, 133–138. [CrossRef] [PubMed]
6. Riley, M.I.; Yoo, W.; Mda, N.Y.; Folk, W.R. Tiny T antigen: An autonomous polyomavirus T antigen amino-terminal domain. *J. Virol.* **1997**, *71*, 6068–6074. [PubMed]
7. Gaudray, P.; Tyndall, C.; Kamen, R.; Cuzin, F. The high affinity binding site on polyoma virus DNA for the viral large-T protein. *Nucleic Acids Res.* **1981**, *9*, 5697–5710. [CrossRef] [PubMed]
8. Pomerantz, B.J.; Mueller, C.R.; Hassell, J.A. Polyomavirus large T antigen binds independently to multiple, unique regions on the viral genome. *J. Virol.* **1983**, *47*, 600–610. [PubMed]
9. Cowie, A.; Kamen, R. Multiple binding sites for polyomavirus large T antigen within regulatory sequences of polyomavirus DNA. *J. Virol.* **1984**, *52*, 750–760. [PubMed]
10. Dilworth, S.; Cowie, A.; Kamen, R.; Griffin, B. DNA binding activity of polyoma virus large tumor antigen. *Proc. Natl. Acad. Sci. USA* **1984**, *81*, 1941–1945. [CrossRef] [PubMed]
11. Cogen, B. Virus-specific early RNA in 3T6 cells infected by a tsA mutant of polyoma virus. *Virology* **1978**, *85*, 222–230. [CrossRef]
12. Farmerie, W.G.; Folk, W.R. Regulation of polyoma-virus transcription by large tumor antigen. *Proc. Natl. Acad. Sci. USA* **1984**, *81*, 6919–6923. [CrossRef] [PubMed]
13. Liu, Z.; Carmichael, G.G. Polyoma virus early-late switch: Regulation of late RNA accumulation by DNA replication. *Proc. Natl. Acad. Sci. USA* **1993**, *90*, 8494–8498. [CrossRef] [PubMed]
14. Cahill, K.B.; Roome, A.J.; Carmichael, G.G. Replication-dependent transactivation of the polyomavirus late promoter. *J. Virol.* **1990**, *64*, 992–1001. [PubMed]
15. Cheng, J.; DeCaprio, J.A.; Fluck, M.M.; Schaffhausen, B.S. Cellular transformation by Simian Virus 40 and Murine Polyoma Virus T antigens. *Semin. Cancer Biol.* **2009**, *19*, 218–228. [CrossRef] [PubMed]
16. Gu, R.; Zhang, Z.; DeCerbo, J.N.; Carmichael, G.G. Gene regulation by sense-antisense overlap of polyadenylation signals. *RNA* **2009**, *15*, 1154–1163. [CrossRef] [PubMed]
17. Beard, P.; Acheson, N.H.; Maxwell, I.H. Strand-specific transcription of polyoma virus DNA early in productive infection and in transformed cells. *J. Virol.* **1976**, *17*, 20–26.
18. Piper, P. Polyoma virus transcription early during productive infection of mouse 3T6 cells. *J. Mol. Biol.* **1979**, *131*, 399–407. [CrossRef]
19. Hyde-DeRuyscher, R.; Carmichael, G.G. Polyomavirus early-late switch is not regulated at the level of transcription initiation and is associated with changes in RNA processing. *Proc. Natl. Acad. Sci. USA* **1988**, *85*, 8993–8997. [CrossRef] [PubMed]
20. Garren, S.B.; Kondaveeti, Y.; Duff, M.O.; Carmichael, G.G. Global Analysis of Mouse Polyomavirus Infection Reveals Dynamic Regulation of Viral and Host Gene Expression and Promiscuous Viral RNA Editing. *PLoS Pathog.* **2015**, *11*, e1005166. [CrossRef] [PubMed]
21. Heiser, W.C.; Eckhart, W. Polyoma virus early and late mRNAs in productively infected mouse 3T6 cells. *J. Virol.* **1982**, *44*, 175–188. [PubMed]
22. Kamen, R.; Jat, P.; Treisman, R.; Favaloro, J.; Folk, W.R. 5' Termini of polyoma virus early region transcripts synthesized in vivo by wild-type virus and viable deletion mutants. *J. Mol. Biol.* **1982**, *159*, 189–224. [CrossRef]
23. Hyde-DeRuyscher, R.P.; Carmichael, G.G. Polyomavirus late pre-mRNA processing: DNA-replication-associated changes in leader exon multiplicity suggest a role for leader-to-leader splicing in the early-late switch. *J. Virol.* **1990**, *64*, 5823–5832. [PubMed]

24. Liu, Z.; Batt, D.B.; Carmichael, G.G. Targeted nuclear antisense RNA mimics natural antisense-induced degradation of polyoma virus early RNA. *Proc. Natl. Acad. Sci. USA* **1994**, *91*, 4258–4262. [CrossRef] [PubMed]

25. Sullivan, C.S.; Sung, C.K.; Pack, C.D.; Grundhoff, A.; Lukacher, A.E.; Benjamin, T.L.; Ganem, D. Murine Polyomavirus encodes a microRNA that cleaves early RNA transcripts but is not essential for experimental infection. *Virology* **2009**, *387*, 157–167. [CrossRef] [PubMed]

26. Adami, G.R.; Marlor, C.W.; Barrett, N.L.; Carmichael, G.G. Leader-to-leader splicing is required for efficient production and accumulation of polyomavirus late mRNAs. *J. Virol.* **1989**, *63*, 85–93. [PubMed]

27. Di Giammartino, D.C.; Nishida, K.; Manley, J.L. Mechanisms and consequences of alternative polyadenylation. *Mol. Cell* **2011**, *43*, 853–866. [CrossRef] [PubMed]

28. Gruber, A.R.; Martin, G.; Keller, W.; Zavolan, M. Means to an end: Mechanisms of alternative polyadenylation of messenger RNA precursors. *Wiley Interdiscip. Rev. RNA* **2014**, *5*, 183–196. [CrossRef] [PubMed]

29. Acheson, N.; Buetti, E.; Scherrer, K.; Weil, R. Transcription of the polyoma virus genome: Synthesis and cleavage of giant late polyoma specific RNA. *Proc. Natl. Acad. Sci. USA* **1971**, *68*, 2231–2235. [CrossRef] [PubMed]

30. Acheson, N. Transcription during productive infection with polyoma virus and SV40. *Cell* **1976**, *8*, 1–12. [CrossRef]

31. Birg, F.; Favaloro, J.; Kamen, R. Analysis of polyoma viral nuclear RNA by miniblot hybridization. *Proc. Natl. Acad. Sci. USA* **1977**, *74*, 3138–3142. [CrossRef] [PubMed]

32. Acheson, N.H. Polyoma giant RNAs contain tandem repeats of the nucleotide sequence of the entire viral genome. *Proc. Natl. Acad. Sci. USA* **1978**, *75*, 4754–4758. [CrossRef] [PubMed]

33. Treisman, R. Characterization of polyoma late mRNA leader sequences by molecular cloning and DNA sequence analysis. *Nucl. Acids Res.* **1980**, *8*, 4867–4888. [CrossRef] [PubMed]

34. Treisman, R.; Kamen, R. Structure of polyoma virus late nuclear RNA. *J. Mol. Biol.* **1981**, *148*, 273–301. [CrossRef]

35. Luo, Y.; Carmichael, G.G. Splice site skipping in polyomavirus late pre-mRNA processing. *J. Virol.* **1991**, *65*, 6637–6644. [PubMed]

36. Acheson, N. Kinetics and efficiency of polyadenylation of late polyomavirus nuclear RNA: Generation of oligomeric polyadenylated RNAs and their processing into mRNA. *Mol. Cell. Biol.* **1984**, *4*, 722–729. [CrossRef] [PubMed]

37. Sung, C.K.; Yim, H.; Andrews, E.; Benjamin, T.L. A mouse polyomavirus-encoded microRNA targets the cellular apoptosis pathway through Smad2 inhibition. *Virology* **2014**, *468–470*, 57–62. [CrossRef] [PubMed]

38. Luo, Y.; Carmichael, G.G. Splice site choice in a complex transcription unit containing multiple inefficient polyadenylation signals. *Mol. Cell. Biol.* **1991**, *11*, 5291–5300. [CrossRef] [PubMed]

39. Huang, Y.; Carmichael, G.G. A suboptimal 5′ splice site is a cis-acting determinant of nuclear export of polyomavirus late mRNAs. *Mol. Cell. Biol.* **1996**, *16*, 6046–6054. [CrossRef] [PubMed]

40. Bass, B.L. RNA editing by adenosine deaminases that act on RNA. *Annu. Rev. Biochem.* **2002**, *71*, 817–846. [CrossRef] [PubMed]

41. Kumar, M.; Carmichael, G.G. Nuclear antisense RNA induces extensive adenosine modifications and nuclear retention of target transcripts. *Proc. Natl. Acad. Sci. USA* **1997**, *94*, 3542–3547. [CrossRef] [PubMed]

42. Gu, R.; Zhang, Z.; Carmichael, G.G. How a small DNA virus uses dsRNA but not RNAi to regulate its life cycle. *Cold Spring Harb. Symp. Quant. Biol.* **2006**, *71*, 293–299. [CrossRef] [PubMed]

43. George, C.X.; Samuel, C.E. Host response to polyomavirus infection is modulated by RNA adenosine deaminase ADAR1 but not by ADAR2. *J. Virol.* **2011**, *85*, 8338–8347. [CrossRef] [PubMed]

44. Kern, F.G.; Pellegrini, S.; Cowie, A.; Basilico, C. Regulation of polyomavirus late promoter activity by viral early proteins. *J. Virol.* **1986**, *60*, 275–285. [PubMed]

45. Zhang, Z.; Carmichael, G.G. The fate of dsRNA in the nucleus. A p54(nrb)-containing complex mediates the nuclear retention of promiscuously A-to-I edited RNAs. *Cell* **2001**, *106*, 465–475. [CrossRef]

46. Chen, L.L.; Carmichael, G.G. Gene regulation by SINES and inosines: Biological consequences of A-to-I editing of Alu element inverted repeats. *Cell Cycle* **2008**, *7*, 3294–3301. [CrossRef] [PubMed]

47. Chen, L.L.; Carmichael, G.G. Altered nuclear retention of mRNAs containing inverted repeats in human embryonic stem cells: Functional role of a nuclear noncoding RNA. *Mol. Cell* **2009**, *35*, 467–478. [CrossRef] [PubMed]

48. Fenton, R.G.; Basilico, C. Changes in the topography of early region transcription during polyomavirus lytic infection. *Proc. Natl. Acad. Sci. USA* **1982**, *79*, 7142–7146. [CrossRef] [PubMed]

49. Ghosh, P.K.; Lebowitz, P. Simian virus early mRNA's contain multiple 5′ termini upstream and downstream from a Hogness-Goldberg sequence; a shift in 5′ termini during the lytic cycle mediated by large T antigen. *J. Virol.* **1981**, *40*, 224–240. [PubMed]

50. Khalili, K.; Feigenbaum, L.; Khoury, G. Evidence for a shift in 5′-termini of early viral RNA during the lytic cycle of JC virus. *Virology* **1987**, *158*, 469–472. [CrossRef]

51. Carbone, M.; Hauser, J.; Carty, M.P.; Rundell, K.; Dixon, K.; Levine, A.S. Simian virus-40 (SV40) small T-antigen inhibits SV40 DNA replication in vitro. *J. Virol.* **1992**, *66*, 1804–1808. [PubMed]

52. Theiss, J.M.; Gunther, T.; Alawi, M.; Neumann, F.; Tessmer, U.; Fischer, N.; Grundhoff, A. A Comprehensive Analysis of Replicating Merkel Cell Polyomavirus Genomes Delineates the Viral Transcription Program and Suggests a Role for mcv-miR-M1 in Episomal Persistence. *PLoS Pathog.* **2015**, *11*, e1004974. [CrossRef] [PubMed]

MDPI AG
St. Alban-Anlage 66
4052 Basel, Switzerland
Tel. +41 61 683 77 34
Fax +41 61 302 89 18
http://www.mdpi.com

Viruses Editorial Office
E-mail: viruses@mdpi.com
http://www.mdpi.com/journal/viruses